P9-EIG-685

Santa Monica Public Library
I SMP 00 1253738 -

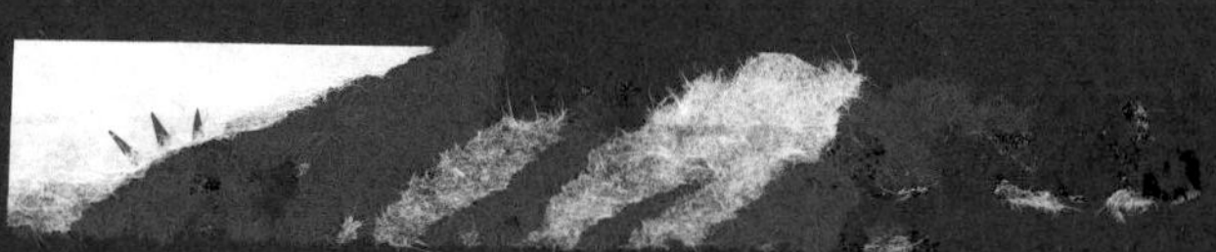

Drawing the Line

Also by Steven M. Wise

Rattling the Cage: Toward Legal Rights for Animals

Acknowledgments

Because I have practiced as an animal protection lawyer for more than twenty years, people often assume that I know a lot about a wide variety of nonhuman animal minds. I really don't. Over the years I have tried to learn by observing, reading, and speaking to scientists who have made the study of animal minds their life's work. The more I learn, the more complex these minds appear, the more I realize I don't know, and the greater respect I have both for the minds of animals and for those trying to figure them out.

When I wrote *Rattling the Cage,* I asked eminent scientists to help me understand the minds of the two species evolutionarily closest to humans, chimpanzees and bonobos. In *Drawing the Line,* I investigated the minds of other species. Again, I asked eminent scientists for help. Again, they responded, discussing cognition, furnishing articles, and reviewing chapter drafts. I owe a great debt to each, and to others.

To film a documentary on my work, Vermont filmmakers Paul Garstki and Donna Thomas generously took me to Kenya and Uganda. Richard Wrangham graciously hosted me for five days at his Ugandan study site in the Kibale Mountains, where Francis Mugurusi, Donor Mukhangyi, Christopher Murvuli, and Maria Llorente tracked chimpanzees for me to observe. Brian Hare and Richard Wrangham reviewed Chapter 1, on chimpanzees.

Martha Nussbaum's 2001 symposium on "Animal Rights" at the University of Chicago Law School provided the forum for me to try out some of the ideas that developed into Chapter 2. Sarah Luick and David Wolfson kindly reviewed Chapter 3. Jean M. Mandler reviewed Chapter 4 on how the cognition of Christopher, my son, developed and explained how Piaget's ideas became anachronisms. Jonas Langer, Sue Taylor Parker, and Anne E. Russon gave me an impromptu seminar on child cognition and explained why Piaget remains important. Michael Tomasello helped me to better understand the whole complicated field and suggested excellent sources of information. John B. Watson provided more articles. M. E. Bitterman reviewed Chapter 5 on honeybee cognition. Bernd Heinrich not only told me about raven minds but, with Irene Pepperberg, reviewed Chapter 6 on Alex the African Grey parrot. Pepperberg introduced me to Alex, Griffin, and Wart at her MIT Lab and answered numerous queries.

Brian Hare, Adam Mklosi, and Dorit Urd Feddersen-Petersen reviewed Chapter 7 on Marbury the dog and provided me with research articles, as did Mary Elizabeth Thurston. Iain Douglas-Hamilton gave me access to his elephant library in Nairobi and, with Cynthia Moss and Joyce Poole, reviewed Chapter 8 about Echo the African elephant. Cynthia Moss generously donated most of a day to introduce me to the elephants of Amboseli National Park in Kenya. Jay Haight, Robert Dale, Terry Maple, Katy Payne, Patricia Simonet, Tara Stoinski, and Betsy Swart provided me with research papers about elephant cognition, helped me find articles, or otherwise aided my search for information about elephant cognition. Martha Nussbaum and Christopher Jones located a story I relate about a Roman elephant. Collins Ajouk, a young Kenyan volunteering at the Sheldrick elephant sanctuary in Nairobi, patiently answered my questions about the orphans he cares for.

Louis M. Herman reviewed Chapter 9 on Ake and Phoenix, two Atlantic bottle-nosed dolphins. Trisha Feurstein was a wealth of information about sources of dolphin cognition. Rick Trout gave me a video on dolphin capture. Kenneth Le Vasseur and Steven Sipman related how they removed two dolphins from Herman's laboratory more than twenty years ago and what has happened since. Ron Schusterman donated an afternoon to talk about dolphin and sea lion cognition and introduced me to his prize student, Rocky the sea lion. Kenneth W. Marten provided me with research about dolphin cognition.

Elizabeth A. Fox, Anne E. Russon, Dan Shillito, Rob Shumaker, Tara Stoinski, and Carel van Schaik reviewed Chapter 10 on Chantek the orangutan, with whom I spent a delightful afternoon in Atlanta in the company of his teacher, Lyn Miles. Michele Goldsmith, Barbara King, Robert W. Mitchell, Penny Patterson, Anne E. Russon, Rob Shumaker, Tara Stoinski, and Joanne Tanner reviewed Chapter 11 on Koko the gorilla. Penny Patterson introduced me to Koko at her famous pupil's home in Woodside, California, and spent an afternoon teaching me about gorillas at the Basel Zoo.

Marc Bekoff, Richard Byrne, Marc Hauser, and Andrew Whiten instructed me on general cognition; and Benjamin Beck, Roger Fouts, Michael Tomasello, and Sue Savage-Rumbaugh helped me unravel the intricacies of great ape cognition.

Antonio Damasio and Jaak Panksepp commented upon my discussion of neuroscience and emotion, while Cam Muir and an evolutionary biologist, who phoned to remain anonymous and whom I refer to as "Mr. Darwin," patiently answered my numerous questions about evolution. Paul Waldau reviewed and commented in detail upon a draft of the finished book. To all these scientists and scholars, so generous with their time, I am enormously grateful.

My agent, Charles Everitt, once again performed in a stellar manner, while my editor, Merloyd Lawrence, continues to embody every writer's dream: intelligent, kind, caring, dedicated to the author's work and to the author, and able cheerfully to suggest vast improvements to a writer's work while making them appear his very own ideas.

Thanks to Gail for everything.

DRAWING THE LINE

Science and the Case for Animal Rights

Steven M. Wise

A Merloyd Lawrence Book

PERSEUS BOOKS
Cambridge, Massachusetts

Cataloging-in-Publication Data is available from the Library of Congress
ISBN 0-7382-0340-8

Perseus Publishing is a member of the Perseus Books Group.
Find us on the World Wide Web at http://www.perseuspublishing.com

Perseus Publishing books are available at special discounts for bulk purchases in the U.S. by corporations, institutions, and other organizations. For more information, please contact the Special Markets Department at the Perseus Books Group, 11 Cambridge Center, Cambridge, MA 02142, or call (800) 255–1514 or (617) 252–5298, or e-mail j.mccrary@perseusbooks.com.

Set in 11.5-point ACaslon by the Perseus Books Group

First printing, April 2002

1 2 3 4 5 6 7 8 9 10—05 04 03 02

For

Roma, Christopher, *and* Siena

No dad ever loved his children more
than your dad loves you

CONTENTS

ONE

In a Ugandan Rain Forest

If humans are entitled to fundamental rights, why not animals? In our considered opinion, legal rights shall not be the exclusive preserve of the humans which has to be extended beyond people thereby dismantling the thick legal wall with humans all on one side and all non-humans on the other side. While the law currently protects wild life and endangered species from extinction, animals are denied rights, an anachronism which must necessarily change.

N. R. Nair v. UOI (Kerala High Court of India, June 6, 2000)

At 6:30 A.M. on June 5, it is scarcely light just fifteen minutes north of the Equator. The rising sun slowly illuminates the ice that tips the massive peaks of Uganda's Ruwenzori Mountains, twenty-five miles east of where I stand. Ptolemy called them the Mountains of the Moon. Ruwenzori National Park borders the Democratic Republic of the Congo; in the summer of 2001, the park is closed. War has ravaged much of the land that runs a thousand miles to the west, Uganda helping to overthrow the present government, which, in turn, overthrew the previous government five years ago. But that's not why the park is closed. Fundamentalist Muslim rebels of the Allied Democratic Front are camped in ravines of the Mountains of the Moon because they are trying to overthrow the Ugandan government, which overthrew its predecessor fifteen years ago. On the drive west, I picked up copies of several Kampalan newspapers. One contained a radio interview with a Muslim rebel commander who was asked why his troops have indiscriminately slaughtered hundreds of Ugandans. He never answered. A couple of pages later I scanned a three-day-old warning from the United States embassy against visiting the area where I was reading the newspaper. Yesterday, I read about a firefight between Ugandan soldiers and rebels trying to escape back to the Mountains of the Moon.

I was leaning against an ebony tree, beginning the third day of a five-day field seminar graciously provided by Richard Wrangham, one of the world's great experts on chimpanzees and, it so happens, the great-great-great-grandson of the famous English abolitionist William Wilberforce. My wife, two independent Vermont filmmakers, Paul Garstki and Donna Thomas, Wrangham, and I were in Kanywara in the Kibale National Forest at the far western end of Uganda. It was our second dawn together.

The first morning, we had risen at 4:30 to "unnest" chimpanzees. That meant following, by flashlight, a native Ugandan tracker, Donor Muhangyi, and a Spanish graduate student, Maria Llorente, who was interested in chimpanzee nestings.

"Watch where you step!"

Terrified of snakes, I played the beam along the forest floor. There was no snake. Instead, I was trespassing in the beanfield that some determined villager had hacked from the steep hillside in what felt to be the middle of a black nowhere.

Donor and Maria finally halted in the pitch dark and said we had reached the place where, the evening before, they had watched chimpanzees ascend and weave their night nests in the trees. We settled onto black plastic garbage bags to protect against the damp ground and silently waited for the chimpanzees to wake. And waited and waited. The long silence was broken only by Wrangham's stage whispers into the blackness.

"Don't point at the chimpanzees. It's not polite."

"Don't eat in front of the chimpanzees. We don't want them to associate us with food."

The apes were sleeping unconscionably late.

"Bunch together. We'll look like a smaller group."

"And if you have to go to the bathroom and leave something solid, go off the trail and dig a hole."

An hour later than usual, Big Brown, a large male, finally peered over the lip of his nest.

Soon, all the chimpanzees were up and out, moving too fast for us to follow. Now in his seventeenth field season of studying the Kibale chimpanzees, Wrangham had begun to understand how they thought. He decided that for breakfast, some of them would head to an old fig tree at the edge of the forest, so we hurried to the tree along the narrow paths Wrangham's team keeps open. When we arrived, sure enough, Big Brown was high in the fig tree, along with his closest ally. No one else showed.

That second morning, we didn't unnest the chimpanzees because they had built their nightly nests in a part of the rain forest nearly impenetrable to humans. Instead, we made for another species of fig tree, its fruit just ripening, to which Wrangham believed the chimpanzees would probably head when they woke up. And that's what they did.

We heard them coming a long way off, screaming, hooting, and drumming on trees. Wrangham told us to bunch together and lower ourselves to appear fewer and less threatening. Suddenly, five males burst into view as they traveled single file along a path that crossed almost in front of us. Each one hesitated when he spotted us, looked us up and down, then hurled himself up into the fig tree.

"That was the core of the male power," Wrangham said.

Eventually, gorged on figs, they took off, with us in slow pursuit through the tangled vegetation and along the narrow trails. By the time we caught up with them, they had killed and eaten two red colobus monkeys and were taking turns dragging the remnants of the carcasses along the ground and into the trees.

"How did the chimpanzees catch them?" I asked Wrangham, for the lithe monkeys were zipping through the trees.

"Sometimes," he said, "by pulling their tails."

Then the chimpanzees were off. And so were we. After five exhausting hours, we struck a fork in the trail and, worn out, quit the pursuit. We decided to rest and refresh ourselves at the edge of the left fork before heading back to camp. Three minutes later, a surprised chimpanzee stumbled onto our group. We looked at him. He looked at us, then took the right fork and disappeared. Wrangham suspected he had been trailing us because he knew we were trailing other chimpanzees and would lead him to them.

Developmental and comparative psychologist Michael Tomasello, who has spent most of two decades studying apes, finds chimpanzees "very sophisticated creatures cognitively."[1] A mountain of evidence supports him. Chimpanzees are probably self-conscious. They use insight, not just trial and error, to solve problems. They have complex mental representations, understand cause and effect, imitate, and cooperate. They compare objects and relationships between objects. They use and make tools. Given appropriate opportunity and motivation, they may teach, deceive, self-medicate, and empathize. They transmit culture between generations.[2]

Captives raised nearly as human have learned thousands of English words at the sophisticated level of a human three-year-old and understand that word order is vital to sentence meaning. In the manner of a human two-year-old they produce hundreds of words and use simple grammar. They point and mentally share the world with humans and other apes. They use symbols in play. They count, perhaps to ten, and add simple numbers and the occasional fraction.[3]

They remember. Ai, a symbol-using chimpanzee, remembers five numbers and their positions for twenty seconds after they disappear from a computer monitor—as good a result, or better, than most human preschoolers.[4] Thirty-four times over nine months, Panzee, a language-trained chimpanzee, watched as objects she desired were hidden outside her enclosure. To obtain them, she had to recruit uninformed humans to help. Up to three days later, Panzee would point toward a hiding place, gesture "HIDE," pant or vocalize while pointing, or use one of 256 abstract keyboard symbols to steer the human where she wanted.[5]

Chimpanzees flourish in rough-and-tumble societies so intensely political and devious they are dubbed "Machiavellian." Their "political intelligence" allows them to make "complicated assessments of power situations," says anthropologist Christopher Boehm.[6] In four chimpanzee societies, two captive, two wild, male and female chimpanzees formed coalitions to subdue the despotic power of an alpha male. Boehm says this action demands such "substantial cognitive ability" that it would be "foolish to deny intentionality where the goal is so unambiguous and the actors are obviously collaborating. It would seem that both wild and captive chimpanzees are able to arrive at and essentially agree upon political strategies, sometimes long-term ones, and shape their societies on that basis."[7]

Wrangham is the thesis adviser of a talented young Harvard anthropology graduate student named Brian Hare, who has spent time in Kibale and understands the natural behavior of chimpanzees. Over lunch in Harvard Square months before, Wrangham told me that Hare has shown that chimpanzees possess elements of a theory of mind of other chimpanzees (the ability to understand and predict another chimpanzee's behavior by attributing mental states).

Hare confirmed what Wrangham told me. With Tomasello and his colleague, Josep Call, Hare devised a way to test whether chimpanzees know what other chimpanzees see and know by exploiting the natural preference of subordinate chimpanzees to avoid competing for food with dominant chimpanzees. When subordinates saw that dominants could see food they could see, they surrendered it to the dominants. But when they saw that a dominant could not see food they could see, they surreptitiously retrieved it, waited until the dominant had gone before consuming it, and sometimes even gave a false signal that kept the dominant away.[8] Hare thought the most interesting finding was that when subordinate chimpanzees in one test became dominants in another test, they acted in the way "you would predict if they are perspective-taking. As dominants they want to monopolize all the food and so they take the 'at risk' piece of food first. Then they take the safely hidden piece."[9] In a second series of experiments, he and his colleagues asked whether chimpanzees know what other chimpanzees know, tested whether one chimpanzee could know what another had, or had not, recently seen, and found they did.[10]

Back to camp. Wrangham opened a box and produced a chimpanzee's "doll." This was one of several "dolls" that Wrangham had acquired since he first saw Kakama playing with a "doll" years ago. Kakama was an eight-year-old whose mother was pregnant when Wrangham spotted him straddling a small log perhaps half his height. For the rest of that morning, Kakama bumped the log behind him, dragged it onto tree nests, played with it in his nest the way a mother would play with a baby, retrieved it when it tumbled thirty feet to the ground, balanced it on his neck, used it as a walking stick, and in general, Wrangham says, "carried that piece of wood in every way imaginable."[11] But Wrangham wasn't able to retrieve the log-doll and told no one what he had seen. Four months later, two of Wrangham's field assistants spotted Kakama doing the same thing and watched for three hours. When he had finished playing, the assistants snatched the log away and labeled it "Kakama's Toy Baby."[12] That was the doll Wrangham showed me. He said they had spotted other chimpanzees playing with other logs in the same way. He had those logs neatly labeled, and filed away, too.

What was an American lawyer doing unnesting chimpanzees in the black predawn of the Kibale mountains? Or examining chimpanzee dolls near where they had been dropped in the forest? Or hurrying along narrow wooded paths to glimpse chimpanzees on the hunt for monkeys? In an earlier book, *Rattling the Cage: Toward Legal Rights for Animals,* I argued that, as the common law stands, normal adult chimpanzees are entitled to basic legal rights because of certain advanced mental abilities they possess. But I never had the privilege of seeing them in the wild, only captive in laboratories and zoos. Now

I could engage in a five-day seminar on chimpanzee behavior and cognition conducted by an expert and confirm for myself that chimpanzees possessed these abilities.

We will look beyond chimpanzees to test the eligibility of normal adult members of seven other species to basic legal rights. I argue that mental abilities that add up to "practical autonomy" are sufficient to entitle any being to basic legal rights. The question is who has them? We'll see how my son, Christopher, mentally developed from an infant not entitled to rights (at least on the ground of autonomy) because of his lack of mental abilities to an entitled child enjoying practical autonomy. Then it's on to nonhuman animals, whom we will compare to Christopher whenever possible.

I didn't have to cross an ocean to meet some of these animals. We'll look, surprisingly, at honeybees, whose brains are a millionth the size of ours, and learn about their unexpectedly complex mental abilities. We'll meet Alex, an African Grey parrot living in Cambridge, Massachusetts. Alex has demonstrated extraordinary mental abilities for an animal with a walnut-sized brain. He can understand, even speak, English words taught him by MIT professor Irene Pepperberg. We'll encounter Marbury, my family dog. He is a member of a species whose mental abilities, again surprisingly for "man's best friend," we have just begun to study formally. We have already learned that dogs read human beings like a book.

I flew to Kenya in search of Echo, an African elephant, and other elephants who live in Amboseli National Park. We'll see that elephants intensely experience a wide variety of emotions, may be self-conscious, have long and accurate memories, communicate, and solve problems. Closer to home, I visited Chantek, a signing orangutan, at his home in Zoo Atlanta. Chantek deceives, pretends, and uses a simple sign language of one hundred and fifty signs, marked by a rudimentary grammar. I called on Koko, a world-famous lowland gorilla, at her home near San Francisco, California. Koko has repeatedly scored between 70 and 95 on standard human child intelligence tests, routinely uses hundreds of signs to communicate, and understands thousands of English words. Both Chantek and Koko easily pass the standard test for self-awareness. Occasionally a door did not open. I was unable to obtain permission to meet Phoenix and Ake, two Atlantic bottle-nosed dolphins, from Louis Herman, the professor who has taught them a simple artificial language at Kewalo Basin in Honolulu, Hawaii.

I really wanted to speak to Herman, for it isn't just the nonhuman animals I met who make this book what it is but the scientists who have spent years working with them; and not just Herman but Pepperberg, who has patiently probed Alex's mind for twenty-five years, Cynthia Moss and Joyce Poole, who know Echo and her family intimately, Lyn Miles and Penny Patterson, who raised and taught Chantek and Koko sign language from her earliest infancy, and Brian Hare, who is just begining to tell us what dogs might think. A glance at the acknowledgments will reveal that I have communicated by letter, e-mail, telephone, and often face-to-face with nearly five dozen of the world's foremost experts in human, ape, dolphin, parrot, dog, elephant, and honeybee cognition, from four continents: Asia, Africa, Europe, and North America. What you will read about the minds of the animals we meet is a product of these scientists' generous teaching and criticisms.

Based on the present state of scientific knowledge about the minds of these animals, I will argue that the case for legal rights for some of them is overwhelming; for others, currently not. But each determination will be saturated in the highest legal values and principles, free of the pervasive legal bias against nonhuman animals, and deeply anchored in scientific fact. To deny the most deserving amongst nonhuman animals basic rights is arbitrary, biased, and therefore unjust. It undermines, and finally destroys, every rationale for basic human rights as well. And states without justice, wrote St. Augustine, are nothing but robber bands.[13]

TWO

One Step at a Time

Obstacles to Ending Nonhuman Animal Slavery

An advocate for the legal rights for nonhuman animals must proceed one step at a time, for progress is impeded by physical, economic, political, religious, historical, legal, and psychological obstacles. Although the historical and legal obstacles remain about the same for every nonhuman of every species, we'll see that the physical, economic, political, religious, historical, legal, and psychological obstacles can loom higher for some nonhumans than for others.

The Physical Problem

The physical problem, that is, the sheer number of nonhuman animals whom we kill, is beyond understanding. More than 300 mammals and birds are killed each time your heart beats. The number triples for the rest of the world. In the United States, more than 10 billion are slaughtered annually just for food. Tens of millions are annually consumed in biomedical research, hundreds of millions more by hunting and entertainment, for clothing, fur, leather, and through numerous other human activities.

The Economic Problem

Political scientist William Lee Miller wrote, "Suppose today some dominant industry, built into the lives and fortunes of a great many people,—to a degree of the whole nation—were found to be morally repugnant; what difficulties there would be in extracting it from the nation's life!"[1] Closely connected to the physical problem is the economic one, for huge primary and numerous secondary industries, generating vast sums, are involved in the raising, using, processing, and killing of those billions of nonhuman animals. Together, the physical and economic obstacles are huge.

Miller was writing of the enslavement of blacks in the antebellum United States. This human slave trade could be immensely profitable.[2] Slavery historian Davis Brion Davis has set out some of those details:

> After decades of research, historians are only now beginning to grasp the complex interdependence of a society enmeshed in slavery. There were shifting interactions among West African enslavers, sellers and European buyers: European investors on the slave trade, who ranged from small-town merchants to well-known figures like the philosophers John Locke and Voltaire: wealthy Virginian and Brazilian middlemen who purchased large numbers of Africans off the slave ships to sell to planters; New Englanders who shipped foodstuffs, timber, shoes and clothing as supplies for slaves in the South and the West Indies; and finally, the European and American consumers of slave-produced sugar, rum, rice, cotton, tobacco, indigo (for dyes), hemp (for rope-making) and other goods.[3]

The interdependence of present society enmeshed in the use of nonhuman animals dwarfs nineteenth-century slave society. Then, it was possible to avoid complicity in human slavery; today the use of nonhuman animal products is so diverse and widespread that it is impossible to live in modern society and not support the nonhuman animal industry directly. For example, the blood of a slaughtered cow is used to manufacture plywood adhesives, fertilizer, fire extinguisher foam, and dyes. Her fat helps make plastic, tires, crayons, cosmetics, lubricants, soaps, detergents, cough syrup, contraceptive jellies and creams, ink, shaving cream, fabric softeners, synthetic rubber, jet engine lubricants, textiles, corrosion inhibitors, and metal-machining lubricants. Her collagen is found in pie crusts, yogurts, matches, banknotes, paper, and cardboard glue; her intestines are used in strings for musical instruments and racquets; her bones in charcoal ash for refining sugar, in ceramics, and cleaning and polishing compounds. Medical and scientific uses abound. And there is much much more.[4]

What Miller concluded in the context of human slavery then is true of nonhuman animals now: "When a pecuniary interest has that magnitude, it is a formidable opponent indeed. Rationalizations are supplied, positions are softened, conflict is avoided, compromises are sought, careers are protected, life goes on."[5]

The Political Problem

In 1999, law professor Richard Epstein insisted "[t]here would be nothing left of human society if we treated animals not as property but as independent holders of rights."[6] This helps explain why the political problem is closely related to the physical and economic problems. Epstein exaggerates: plenty would be left to human society. But the massive industries that have developed around the exploitation of nonhuman animals would be damaged, perhaps destroyed. Certainly industries that depend upon inflicting bodily harm upon nonhuman animals, such as the meat or biomedical research industries, would be severely affected if the animals upon which they depend were given a legal right to bodily integrity.

Turn again to human slavery for a comparison. Britain's eighteenth- and nineteenth-century slave trade centered on newly enriched Liverpool, in the bottoms of whose ships traveled more than half of Europe's slaves.[7] Alarmed by rising abolitionist voices, Liverpudlians composed a ditty:

> If our slave trade be gone, there's an end to our lives,
> Beggars all we must be, our children and wives;
> No ships from our ports their proud sails e'er would spread,
> And our streets grown with grass, where the cows might be fed.[8]

Hand in glove with the West Indian planters, Liverpool's members and witnesses argued in Parliament that their slavers were men "of impeccable characters," that the trade itself was not cruel. The Middle Passage was "one of the happiest periods of a Negro's life." It was crazy to think men "whose profit depended on the health" of slaves might harm them. The slavers did the Africans a favor; if they "could not be sold as slaves, they would be butchered and executed at home." Abolition would ruin Britain and her colonies, throw thousands of sailors out of work, destroy her seafaring training ground. It would cripple British seapower. And all would be for naught, for if Britain abolished its slave trade, other nations would fill the vacuum and steal Britain's profits. Abolition would send many of Britain's "most loyal, industrious, and useful subjects" emigrating to America.[9]

Liverpool's sailmakers, bakers, and gunmakers petitioned Parliament to keep slavers sailing. To whom would they sell their sails, biscuits, and guns?[10] Abolitionists were advised to direct their energies toward alleviating the plight of poor young chimney sweeps.[11] Doomsayers were not confined to Britain: in 1832, the future president of the College of William and Mary (my alma mater), Thomas Roderick Dew, wrote, "It is in truth the slave labour in Virginia which gives value to her soil and her habitations . . . take away this and you pull down the atlas that upholds the whole system . . . eject from the state the whole slave population . . . and the Old Dominion will be a 'waste howling wilderness,' . . . 'the grass shall be seen growing in the streets, and the foxes peeping from their holes.'"[12]

In Britain, Parliamentary fellow travelers argued slaving was "not an amiable trade but neither was the trade of a butcher an amiable trade, and yet a mutton chop was, nevertheless, a good thing." Abolition would destroy Africa. One Member of Parliament rhapsodized that slaves appeared so happy he often wished to be one.[13]

On May 12, 1789, William Wilberforce, who had long hoped to strike a blow at the slave trade and who would be instrumental in smashing it eighteen years later, delivered a famous speech in the House of Commons. The speech included a listing of woes similar to those predicted by the Liverpudlians; Wilberforce then disclosed that it had been written before the American Revolution, predicting the economic ruin of Britain should it lose its colonies.[14] William Pitt, the prime minister, summed the slaving arguments: "The blood of these poor negroes was to continue flowing; it was dangerous to stop it because it had run so long; besides, we were under contract with certain surgeons to allow them a certain supply of human bodies every year for them to try experiments on, and this we did out of pure love for science."[15] Britain eventually did the right thing, if not always for the purest of reasons. Within thirty years, this enormous institution was gone.

As Miller points out, slave owners in antebellum United States "were a major and very powerful part of the population, strategically located, from the outset in the highest places."[16] Five of the first seven presidents of the United States owned slaves. For fifty of the first sixty-four years of the life of the nation's country, the president was a slaveholder; for twenty-eight of the country's first thirty-five years, the Speaker of the House of Representatives owned slaves. The president pro tem of the United States Senate was always a slaveholder, as were the majority of every presidential cabinet and the U.S. Supreme Court up to the Civil War. George Washington, Thomas Jefferson, Andrew Jackson, and John Marshall all owned slaves.

Some readers may shift uncomfortably at comparisons between human and nonhuman slavery. I began *Rattling the Cage* by recalling the brutish life and lingering death of Jerom, a chimpanzee whom biomedical researchers imprisoned for life inside a small, dim, often chilly cell that lay within a large windowless grey concrete box at the Yerkes Regional Primate Research Center in Atlanta, Georgia. Without mercy, and from the time he was a baby, they repeatedly infected Jerom with HIV viruses. After a hellish decade, he died. In a February 2000 speech in Boston's Faneuil Hall, constitutional law professor Laurence Tribe said, "Clearly, Jerom was enslaved."[17]

The first definition of "slave" in the *Oxford English Dictionary* is "[o]ne who is the property of, and entirely subject to, another person, whether by capture, purchase, or birth: a servant completely divested of freedom and personal rights."[18] International law has, for most of a century, defined slavery as "the status or condition of a person over whom any or all of the powers attaching to the right of ownership are exercised."[19]

Philosopher Isaiah Berlin claims that when a human stops me from doing what I want to do, I've lost some liberty. When it has been reduced "beyond a certain minimum, I can be described as being . . . enslaved."[20] That student of human slavery, sociologist Orlando Patterson, characterizes a slave as "the ultimate human tool, as imprintable and as disposable as the master wished."[21] David Brion Davis argues slavery is just a branch of property law that "starts with the power of one individual over another."[22]

There are different kinds of slaves. Law professor Alan Watson finds a three-tier hierarchy of animal slavery has long existed in the West. At the top are humans enslaved for reasons other than race: Persians enslaved by classical Greeks, Moslems by Renaissance Christians, Christians by Moslems. Below these are humans enslaved because of their race: Africans enslaved by Europeans and Americans. At the bottom are nonhuman animals, enslaved by everyone.[23] What all have in common is their legal thinghood.

For most of history, nobody cared about human slavery. Edith Hamilton wrote of Athenian slaves: "Everyone used them; no one paid attention to them."[24] Two thousand years later, "refuse slaves," the sick and the weak from the Middle Passage, the unsold, were often abandoned on American quays to die. No one cared.[25] Here is what Charles Darwin saw on his visit to Brazil:

> I lived opposite to an old lady who kept screws to crush the fingers of her female slaves. I have stayed in a house where a young household mulatto, daily and hourly, was reviled, beaten, and persecuted enough to break the spirit of the lowest animal. I have seen a little boy, six or seven years old, struck thrice with a horse-whip (before I could interfere) on his naked head, for having handed me a glass not quite clean

> . . . I was present when a kind-hearted man was on the point of separating forever the men, women, and children of a large number of families who had long lived together.[26]

"What must be remembered," Hamilton said, "is that the Greeks were the first who thought about slavery. To think about it was to condemn it."[27] The freedom that gripped Greece slowly diffused through the West and the rest of world.[28] But it took more than 2,000 years of thinking about human slavery before a sufficiently large number of free men and women hated it enough to kill it. Darwin wrote in his journal that "[i]t is argued that self-interest will prevent excessive cruelty; as if self-interest protected our domestic animals."[29] Richard Sorabji, professor of ancient philosophy, has concluded "the modern debate on the treatment of animals has, in fact, reached the same point as the ancient debate on slavery in Aristotle's time."[30]

Of course, human slavery apologists tried to distinguish slavery from tyranny.[31] Of the Atlantic slave trade, Orlando Patterson writes, "The strangeness and seeming savagery of the Africans, reinforced by traditional attitudes . . . 'were major components in that sense of *difference* which provided the mental margin absolutely requisite for placing the European on the deck of the slave ship and the Negro in the hold.'"[32] The ancient Great Chain of Being, by which the universe was imagined as a divinely ordained hierarchy, proved indispensable. The most famous fugitive slave of all save Spartacus, the American Frederick Douglass, wrote that when his dead master's property was inventoried, livestock and slaves were valued as one, "horses and men, cattle and women, pigs and children, all holding the same rank in the scale of being."[33]

As legal things, nonhuman animals are treated today as human slaves were treated once and continue to be treated in those few places in which human slavery is unlawfully practiced. The African American writer Alice Walker says that this, "even for those of us who recognize its validity, is a difficult one to face. Especially so if we are the descendants of slaves. Or of slave owners. Or of both. Especially so if we are also responsible in some way for the present treatment of animals [or] . . . if we are complicit in their enslavement and destruction, which is to say at this juncture in history, master."[34]

Sometimes the intersection between human and nonhuman animal slavery startles. The teeth of African elephants, ivory tusks, were once used in great quantity for piano keys, billiard balls, combs, page markers, letter openers, business cards, domino pieces, and other essentials. One obtains tusks by killing elephants and sawing the tusks from the carcasses. As Douglas Chadwick explained in *The Fate of the Elephant*, hundreds of years ago,

> tusks were what traders wanted most: enslaving people was merely the most expedient way to transport the teeth great distances through trackless countryside. Arab traders who continued to control the main trading operations, sometimes captured slaves only to ransom them for ivory. Then they would snatch more slaves to haul the tusks out. Once at the coast, both tusk and transporter were sold, the tusk typically fetching at least twice as much as the person who has carried it.[35]

Writers often compare human slaves to nonhuman animals. "The truly striking fact about slavery," David Brion Davis writes, is the "antiquity and almost universal acceptance of the *concept* of the slave as a human being who is legally owned, used, sold, or otherwise disposed of as if he or she were a domestic animal. This parallel persisted in the similarity of naming, branding, and even pricing slaves according to their equivalent in cows, camels, pigs, and chickens."[36] Legal scholar Roscoe Pound said that in Rome a slave "was a thing, and as such, like animals could be the object of rights of property."[37] Barry Nichols, an expert in Roman law, says in Rome, "the slave was a thing . . . he himself had no rights: he was merely an object of rights, like an animal."[38] Rome's initial regulation of the treatment of slaves "took the same form as our legislation for the protection of animals. The master might be punished criminally for abuse of his powers, but the slave could not himself invoke the protection of the law."[39] This was true in slaveholding Virginia.[40] And Mississippi.[41]

It is therefore no coincidence that as the first fetters were being struck from blacks in England in the eighteenth century, a sustained movement toward justice for nonhuman animals began. Some of the same most prominent nineteenth-century antislavery British parliamentary campaigners, including William Wilberforce and Foxwell Buxton, provided the earliest support for legislation against cruelty to nonhuman animals and helped create the world's first permanent organization dedicated to preventing this cruelty, The Royal Society for the Prevention of Cruelty to Animals.[42] Across the Atlantic, the founders of the American Society for the Prevention of Cruelty to Animals in New York City helped organize the New York Society for the Prevention of Cruelty to Children.[43] The American feminist movement sprang from and often remained intertwined with abolitionism.[44] Lucy Stone, Amelia Bloomer, Susan B. Anthony, and Elizabeth Cady Stanton would meet with the antislavery newspaper editor Horace Greeley to toast "Women's Rights and vegetarianism."[45]

All this partially explains why, though dozens of slave societies have existed, I often return to the Atlantic slave trade and British and American slavery when I discuss the plight of nonhuman animals. Human slavery was made possible by the legal rule that humans could be legal things, a concept that seems so wrong today, yet was deeply woven into the societies of its day. The arguments in its favor are often repeated by those who, as Alice Walker says, are "complicit" in the enslavement of nonhuman animals.[46]

The Religious Problem

Genesis says God granted humans dominion "over the fish of the sea, and over the fowl of the air, and every living thing that moveth upon the earth."[47] Both the Old and New Testaments, the apostle Paul, St. Augustine, and St. Thomas Aquinas stitched into the fabric of Judeo-Christian doctrine the idea that nonhuman animals had been created for the benefit of humans.[48] The twentieth-century Christian theologian and Oxford don C. S. Lewis thought a Christian vivisectionist might therefore justify torturing a chimpanzee to death on two grounds. She might admit she was a speciesist. This is why Robert Speth, professor of pharmacology and neuroscience at Washington State University, argues that chimpanzees and bonobos can never have legal rights: "Apes are not humans" and "Humanity is our species and . . . our first and primary obligation is to ourselves."[49] Or he might declare that humans are preeminent, created in the image of God, therefore superior to every other creature and follow God's ordering of the universe. This depends not a whit upon either evidence or reason and can be neither proved nor disproved.

Religion, however, need not obstruct animals' rights. That it often does should come as no surprise; after all, religion has often been indifferent, even hostile, toward human rights.[50] The major Western religions, formed centuries before human rights appeared, long ignored them. The Atlantic slave trade was thoroughly dominated by Christians. Jews enslaved gentiles, Africans, and other Jews.[51] Islam accepted slavery without question, and its slave trade may have equaled the Atlantic slave trade.[52] American slave supporters trumpeted God's imprimateur on slavery from Genesis to St. Paul. Before the American Civil War, Baptist minister Thornton Stringfellow argued that the institution of slavery had received "the sanction of the Almighty" and that the "control of the black race, by the white, is an indispensable Christian duty."[53] Even the war itself did not strangle the claims of some Orthodox Jews and Christian biblical scholars that human slavery was "an ordinance of God."[54] Similarly, Judeo-Christian theologians long argued that women were supposed to be subjugated by men.[55] Of the common law's subjugation of married women to their husbands, one seventeenth-century English legal scholar thought "[t]he common laws here shaketh hand with divinitye."[56]

Religious faith can blind adherents to facts. Listen to the verdict of Galileo's inquisitors: "We say, pronounce, sentence, and declare that you, Galilieo . . . have rendered yourself in the judgment of this Holy Office vehemently suspected of heresy, namely of having held and believed the doctrine which is false and contrary to the Sacred Scriptures that . . . the earth moves and is not the center of the world."[57] In the twenty-first century, a member of the Kansas Board of Education, who had just voted to remove both Darwinian evolution and physics' Big Bang theory from school curriculums, said, "I don't believe that humans descended from apes, no. How come there's still apes running around loose and there are humans? Why did some of them decide to evolve and some did not?"[58]

Evolutionary biologist Ernst Mayr has written, "[n]o educated person any longer questions the validity of the so-called theory of evolution, which we now know to be a simple fact."[59] On the cusp of the June 2000 announcement that the human genome had been pieced together, David Baltimore, president of the California Institute of Technology and winner of the Nobel prize in medicine, wrote, "it confirms something obvious and expected, yet controversial: our genes look much like those of fruit flies, worms, and even plants . . . the genome shows that we all descended from the same humble beginnings and that the connections are written in our genes. That should be, but won't be, the end of creationism."[60]

Judges freely admit that religious beliefs may be neither logical nor consistent.[61] Judges must respect the right of all to believe what they wish. But secular law should soar beyond belief. A human-nonhuman animal hierarchy constructed before science was born, before rights were invented, and for which no objective evidence exists, one that allows a believer to do as he pleases with Creation seems a bit unfair to Creation. You can believe in hierarchy. You can believe the universe was made just for you. Hierarchy and a major sense of entitlement are not insurmountable problems. The problem occurs when you treat as slaves those whom you believe lie beneath you. Religion once sustained human slavery. It was wrong then. When it blindly sanctions the slavery of every nonhuman animal, it is wrong now.

The Historical Problem

The idea that everything exists for the sake of humans was a core belief of the highly influential ancient Stoics, first in Greece, then in

Rome; the same idea was also found in Old Testament Law Codes and other ancient law. In the fifth century, St. Augustine, who thought Christ a Stoic in the way he viewed nonhuman animals, fused the animal-related teachings of the Hebrews and the Stoics and folded them into Christianity.[62] A century later, Byzantine Emperor Justinian injected the same teachings into his immensely influential legal codes. From there, they were absorbed into the legal writings of Continental Europe, taken up by the great lawyers, judges, and commentators of English common law, and received nearly whole in America.[63] That is why this idea remains at the root of what the law says we in the West can do to nonhuman animals today.[64]

Most of the highly influential ancient Stoics thought animals had life, sensation, and impulse, but lacked emotions, reason, belief, intentionality, thought, and memory.[65] The Roman Stoic, Seneca, thought nonhuman animals could grasp only what they sensed. In his stable, Seneca said, a horse "is reminded of the road when it is brought to where it starts. But in the stable it has no memory of it however often it has trodden. As for the third time, the future, that does not concern dumb animals."[66]

Many nonhuman animals probably live just in the world they sense, just as Seneca thought. Every human, when an infant, probably lives in that perceptual world, though scientists don't agree for how long. Thousands of humans never develop beyond it, and thousands return to such a state in old age. But millions of nonhuman animals also live at least partially in conceptual worlds, unchained to the present, unrestricted to the experiences of their senses, able to think.

A designed universe formed the core of the immensely influential "Great Chain of Being." Rational humans occupied the topmost rungs reserved for corporeal creatures. Lesser Creation was aligned below them. It wasn't just that nonhuman animals occupied lower rungs, but that we differed so fundamentally from them that we were incomparable.[67] Nonhumans literally were made for us: savage beasts foster our courage and train us for war; singing birds exist to entertain us; cows and sheep provide us with fresh meat; lobsters feed us and provide us with exercise as we crack their shells—which also double as nifty models for body armor; and lice make us adopt clean habits.[68]

C. S. Lewis thought "[t]he only rational line for the Christian vivisectionist to take is to say that the superiority of man over beast is a real objective fact, guaranteed by Revelation, and that the propriety of sacrificing beast to man is a logical consequence." The world conforms "to a hierarchical order created by God and is really present in the universe whether any one acknowledges it or not."[69]

The Legal Problem

Generally, law divides the physical universe into persons and things. Things are objects over which a person exercises a legal right.[70] Roscoe Pound called legal persons "the unit of the legal order."[71] A button, for example, buys nothing; the dollar does. For all of Western human history, nonhuman animals have been buttons in the legal system; so were human slaves and women and children.[72]

Those who write about persons and things may struggle to distinguish them. One of the most prominent legal scholars ever to write, John Austin, defined things as "such permanent objects, not being persons, as are sensible or perceptible through the senses."[73] Not helpful. Human slaves, "[l]ike cattle . . . are things and the object of rights; not persons and the subjects of them."[74]

Daniel Defoe wrote:

> *Nature has left this tincture in the blood,*
> *That all men would be tyrants if they could.*[75]

Humans can freely be tyrants over things. Personhood is the legal shield that protects against human tyranny; without it, one is helpless. Legally, persons count, things don't.[76] Until, and unless, a nonhuman animal becomes a legal person, she will remain invisible to civil law. She will not count.

The Psychological Problem

Finally, there is the psychological problem. Millions believe that nonhumans lack every important mental ability, that they are made for humans, that this was how the universe was designed; for these reasons, they can legitimately be viewed as things, not persons, in law. Beliefs are personal, subjective, often unproveable and illogical. They dictate what we think is possible.[77] What we see may depend upon what we believe *can be* seen.[78] Law ignores the impossible. For centuries, lawyers and judges have believed that nonhumans could not possibly have rights.

Sophisticated judges know they decide cases under the influence of their beliefs. "When I say that a thing is true," declared the most famous American judge, Oliver Wendell Holmes, Jr., "I mean that I cannot help believing it."[79] Moral judgments are influenced by intuition, experience, and emotions.[80] Judges approach their work with different visions of law and justice constructed from their different beliefs hammered in the unique crucible that is each of their lives.[81] For Holmes, some beliefs are kin to liking beer.[82] Perhaps one man "cannot help" but believe that he can enslave others of another race, sex, religion, nationality, or species, while another "cannot help" but believe that he can't. Such beliefs "cannot be argued about," Holmes wrote, "and therefore, when differences are sufficiently far reaching, we try to kill the other man rather than let him have his way."[83] This is the stuff of revolution, scientific, social, political; and Holmes, wounded three times in the American Civil War, knew this firsthand.

Shifts occur only after people come to believe that something is possible. This book argues that at least some nonhuman animals should have basic legal rights. At its core is the supporting scientific evidence, much of which is currently known only to a cadre of experts in scientific subdisciplines. Making the argument is the first step toward informing policymakers, judges, and the public about what is known, and, therefore, attaining the goal.

This obstacle is well illuminated again by human slavery. Slave historian David Brion Davis writes:

> Today it is difficult to understand why slavery was accepted from prebiblical times in virtually every culture and not seriously challenged until the late 1700s. But the institution was so basic that genuine antislavery attitudes required a profound shift in moral perception. This meant fundamental religious and philosophic changes in views of human abilities, responsibilities and rights.[84]

William Miller, one last time:

> Thinkers and statesmen and leaders and realistic politicians of all stripes believed that American slavery could not be ended—not by deliberate human action. Those who supported slavery belligerently asserted that it could not be done; those who deplored slavery sorrowfully granted that it could not be done; those who had unsorted mixtures of opinions—the great majority, let us guess—felt that it could not be done, and did not want to hear about it.[85]

Avoiding Speciesism

"If we cut up beasts simply because they cannot prevent us and we are backing our own side in the struggle for existence, it is only logical to cut up imbeciles, enemies, or capitalists for the same reasons."[86] So wrote C. S. Lewis. I began my "Animal Rights Law" class at the Harvard Law School by hanging a sign on the classroom door upon which my daughter, Roma, had scrawled "No pigs." This replicated the sign hung on the doorknob of the Harvard Law School classroom in David Mamet's children's book, *Henrietta.*[87] Henrietta was a pig. Having "aspired to the Law" as a piglet, she applies to Harvard Law School.[88] Summarily rejected, she haunts the law school's lecture halls until barred as a nuisance and the "No Pigs" sign is hung from doorknobs. But Henrietta has the great fortune to aid the university's nearsighted president, who can't see that she's a pig. She so impresses him she is admitted to the law school and ascends to the United States Supreme Court.

Mamet's book is an allegory about racism and sexism and every other "ism" by which humans arbitrarily favor their own kind. It's also about "speciesism." Coined nearly thirty years ago by British psychologist Richard Ryder, "speciesism" is defined by the *Oxford English Dictionary* as "[d]iscrimination against . . . animal species by human beings, based on an assumption of mankind's superiority."[89] In other words, it's a bias, as arbitrary and hateful as any other. The English philosopher R. G. Frey, who opposes rights for nonhuman animals, "cannot think of anything at all compelling that cedes all human life of any quality greater value than animal life of any quality."[90]

Since Linnaeus invented the system of biological classification we use today, humans have assigned themselves to their own "family," *Hominidae,* the taxonomic classification just above "genus," *Homo,* and "species," *sapiens.* We stuck even our closest cousins, the chimpanzees and bonobos, into a separate family. Ethologists Lesley Rogers and Gisela Kaplan point out that some scientists have suggested that at least some apes join us in *Hominidae.* If they did, "[w]e would have reason to extend to the other apes some, if not all, of the rights that we presently afford humans."[91] But what reason would that be? Why should species, genus, and family be relevant to the assignment of legal rights? Shouldn't it be what they are, not who they are, that counts? To avoid speciesism and still justify depriving every nonhuman animal of rights, we must identify some objective, rational, legitimate, and nonarbitrary quality possessed by every *Homo sapiens,* but possessed by no nonhuman, that entitles all of us, but none of them, to basic liberty rights. I shouldn't search too long because this quality doesn't exist. In this chapter, I identify one quality, practical autonomy, that is sufficient to entitle any being of any species to liberty rights.

Taxonomy never determines rights. It can, however, allow quick and often reliable determinations about whether a species' normal adult members might be eligible for basic legal rights once we understand what cognitive capabilities they possess.

I devote a chapter each to honeybees, African Grey parrots, dogs, African elephants, Atlantic bottle-nosed dolphins, orangutans, and gorillas; I focus, with the exception of honeybees, upon an individual. Strictly speaking, individual results are anecdotes. But as comparative psychologist Marc Hauser notes:

> There are anecdotes and there are anecdotes. For example, Jane Goodall provides a detailed account of the murderous attacks by Passion (a chimpanzee) on all of Pom's (another chimpanzee) infants. This is a single case study of chimpanzee brutality, but the details are extensive. In this sense many of the cases of deception from captive apes come from detailed observations of a single animal over years and years of observation. Rejecting these single cases would be like ignoring someone who announced that they had a dog who could write computer code in Pascal, C, and Fortran. One would be foolish to ask for twenty more dogs in order to seal the case shut! One case shows that a capacity exists in the species.[92]

Irene Pepperberg, acclaimed for her work with Alex, the African Grey parrot whom we will meet in Chapter 6, agrees: "If one subject reliably performs a given task at a particular age, the tested aptitude is within the species capacity at that point in development."[93]

One Step at a Time for Legal Rights

The most basic rights of nonhuman animals must be recognized first. What are they? What nonlawyers think of as one legal right is usually a bundle. Law professor Wesley Hohfeld untied these bundles almost a century ago.[94] He thought that when one person has a legal advantage (that's the right), another person bears the corresponding legal disadvantage. Like low- and high-pressure systems on a weather map, neither exists alone, and Hohfeld defined them in relationship to each other.[95]

Hohfeld set out four kinds of legal rights.[96] One is the "liberty"; this lets us do what we please but has little practical value because no one has to respect it. The second is the "claim," which commands respect.

A claim can constrain my liberty because I have a duty to act or not act in certain ways toward someone with a claim.[97] It's not agreed whether we have to be smart enough to assert a claim to have one. We probably don't.[98] But to be safe, I won't argue that nonhuman animals are entitled to claims because probably none is smart enough to assert one. A person can use a "power" to affect another's legal rights, the power to sue being perhaps the most important.[99] It's not clear whether a person has to be smart enough to assert a power to have it—probably not. But to be safe, I won't argue that nonhuman animals are entitled to powers, either. Last, an "immunity" legally disables another person from interfering with you.[100] Claims say what we should not do; immunities what we *cannot* legally do. I may kidnap you, but I can't enslave you because human bondage is prohibited under domestic and international law. Persons are simply *immune* from enslavement.[101] Rational arguments cannot be made that someone must be smart enough to assert an immunity to have one as immunities don't need to be asserted. Such immunities as freedom from slavery and torture are the most basic kinds of legal rights, and so it's these to which nonhuman animals, like human beings, are most strongly entitled.[102]

One Step at a Time for Law

I use a "round hole, square peg" theory of legal change. The round hole is the legal system, which treats nonhuman animals as rightless things. The square peg is the rule that nonhuman animals are persons entitled to basic legal rights. To attain these rights, one must either square the hole or round the peg.

The common law—that jewel of English-speaking jurisprudence—is ideal for peg-rounding. Lemuel Shaw, the most prominent nineteenth-century American common-law judge, said it consists of a "few broad and comprehensive principles, founded on reason, natural justice, and enlightened public policy, modified and adapted to all the circumstances of all the particular cases that fall within it."[103] It can accommodate different visions of what law is, and those visions vary dramatically.

Judges holding "formal visions" inflexibly marinate their decisions in the past. Formal judges think judges should decide the way judges have always decided because judges have always decided that way. The most formal of these judges—I call them "Precedent (Rules) Judges"—think it better that law be certain than right. They see law as a system of narrow and consistent rules they can more or less apply mechanically. These judges aren't being stubborn or lazy or brainless; they value, or they think a legal system should value stability, certainty, and predictability.

Here is a startling example of how a Precedent Rules court dealt with a case involving nonhuman animals. In June 2001, the Wisconsin Supreme Court refused to change the law that forbids an owner from obtaining emotional distress damages when her companion nonhuman is unlawfully killed before her eyes because he is property.[104] A policeman had illegally shot a family's dog. The Court said:

> At the outset, we note that we are uncomfortable with the law's cold characterization of a dog . . . as mere "property." Labeling a dog "property" fails to describe the value human beings place upon the companionship that they enjoy with a dog. A companion dog is not a fungible item, equivalent to other items of personal property. A companion dog is not a living room sofa or dining room furniture. The term inadequately describes the relationship between a human and a dog. . . . Nevertheless, the law categorizes the dog as personal property despite the long relationship between dogs and humans. To the extent this opinion uses the term "property" in describing how humans value the dog they live with, it is done only as a means of applying established legal doctrine to the facts of the case.[105]

These judges did not believe that "property" properly describes a companion nonhuman. Yet because prior judges had believed it, they felt obliged to condone the anachronism. "Precedent (Principles)

Judges" look to a different past. Previous decisions, or precedents, lay down not narrow rules but broad principles, and judges needn't confine themselves to the specific ways in which judges have applied them. If justice demands a rule be changed, judges use these principles to reconstruct the law in ways that might have astounded the earlier judges.

"Substantive Judges" reject the past as manacle. Their legal vision starts with social considerations: moral, economic, political.[106] Law should express a community's present sense of justice, not that of another age.[107] Courts should keep law current with public values, prevailing understandings of justice and morality, and new scientific discoveries. Principles and policies live and die, and when they die, they should be buried. Substantive judges want to know *why* judges once decided a certain way and whether these reasons still make sense. They don't want issues just settled, but decided correctly, and they will change law to get it right.

Substantive judges who try to peer into the future are "Policy Judges." They want to predict the effects of their rulings and think law should be used to achieve such important goals as economic growth, national unity, and the health or welfare of a community. "Principle Judges" value principles and right when deciding cases.[108] They borrow these principles from religion, ethics, economics, politics—almost anywhere—and these may range from representative democracy to the maximization of wealth to the two most important legal principles, liberty ("the supreme value of the Western world") and equality ("far and away the most important").[109] Most judges, especially in the United States, are either "Precedent (Principles) Judges" or "Principle Judges," which every judge should be when confronted with the question of who is entitled to legal personhood.

Meeting the Law on Its Own Terms

Richard Posner, perhaps the leading scholar-judge in the United States, wrote in the *Yale Law Journal* that *Rattling the Cage* "is not an intellectually exciting book. I do not say this in criticism. Remember who Wise is: a practicing lawyer who wants to persuade the legal profession that courts should do more to protect animals."[110] Posner is correct. In arguing for the basic rights of nonhuman animals, I rely upon the long-standing principles of liberty and equality.

Liberty and equality are the first principles of Western law. They are enshrined in the American Declaration of Independence: "We hold these truths to be self-evident, that all men are created equal, that they are endowed by their Creator with certain unalienable Rights, that among these rights are Life, Liberty, and the pursuit of Happiness." French Revolutionaries demanded liberté and égalité. Abraham Lincoln placed them at the beginning of his Gettysburg Address: "Four score and seven years ago, our fathers brought forth upon this continent a new nation, conceived in liberty and dedicated to the proposition that all men are created equal." They are found in Article I of the Universal Declaration of Rights: "All human beings are born free and equal in dignity and rights."

Equality demands likes be treated alike. Equality rights depend upon how one rightless animal compares to another who enjoys rights. Any animal might be entitled to equality rights, even if she isn't entitled to liberty rights, because she is "like" someone who possesses basic liberty rights. We'll take a closer look at which nonhuman animals might be entitled to equality rights in Chapter 12.

Liberty entitles one to be treated a certain way because of how one is made. It doesn't matter how others are treated. Because liberty rights turn on a being's qualities and because a certain degree of autonomy will suffice to entitle one to rights, we will puzzle out which mental abilities the individuals of seven species possess.

Since World War II, nations have agreed that liberty to act as one pleases stops somewhere, though they don't always agree where.[111] Some absolute and irreducible minimum degree of bodily liberty and bodily integrity is everywhere considered sacrosanct. If we trespass upon them we inflict the gravest injustice, for we treat others as slaves and things.[112] And we may not enslave. We may never torture. Yet these sacred principles are the front line in the battle for the rights of nonhuman animals.

An important aspect of liberty is autonomy and self-determination. As C. K. Allen, one prominent legal writer, observes:

> The essential difference between person and thing seems to lie in the quality of *volition.* Animate creatures clearly possess some kind of motive-power which corresponds with the human will; it may be a very strong force indeed—thus if we wish to attribute to a man a particularly obstinate will, we compare him to a mule. But it is not a kind of will which is recognized by law; it cannot in modern societies, involve the creature which exercises it in any consequences of right or liability. Nor do we attribute to the creatures what is closely kin to this volitional capacity, that is, the power of reason. Hence a thing has been defined [as having a] "*volitionless* nature."[113]

Allen implies a mule's will is unrecognized in law because, though purposeful, it comes from instinct, which is the antithesis of volition. The mule has a will. He just can't control it. Whether we call it self-determination, autonomy, or volition, it is sufficient for basic legal rights. Things don't act autonomously. Persons do. Things can't self-determine. Persons can. Things lack volition. Persons don't. Persons have wills.[114] As we explore whether an elephant, dolphin, gorilla, dog, orangutan, parrot, or honeybee is entitled to basic rights, it will be important to see whether we can detect a capacity for autonomy.

Philosophers often understand autonomy, which includes self-determination and volition, the way Immanuel Kant did two centuries ago. I call his "full autonomy." Nonhuman animals, and probably children, act from desire, Kant believed.[115] Fully autonomous beings act completely rationally, and their ability to do that demands they be treated as persons.[116]

Kant is not the only philosopher to try to knit hyperrationality into the fabric of liberty.[117] The most honest concede what philosopher Carl Wellman calls a "monstrous conclusion": a great many human beings don't make the cut.[118] Most normal adults lack it. Infants, children, the severely mentally retarded, autistic, senile, and the persistently vegetative never come close. Were judges to accept full autonomy as prerequisite for personhood, they would have to exclude most humans. Yet these complex autonomies are the sort to which primatologist Frans de Waal refers when he claims that apes, smart and cultured as he knows they are, can never have legal rights. He thinks rights nonsense unless accompanied by responsibilities of a kind that apes, and millions of humans, can never shoulder.[119]

Judges decisively reject Kant's "full autonomy." Events on Leap Day 2000 in the United States show how wrong de Waal and Kant are. A six-year-old Michigan first-grader smuggled a handgun into school and shot a classmate to death. The county prosecutor issued this statement: "There is a presumption in law that a child . . . is not criminally responsible and can't form an intent to kill. Obviously he has done a very terrible thing today, but legally he can't be held criminally responsible."[120] The child couldn't successfully be sued civilly, either.

Isaiah Berlin explained that, for Kant, "[f]reedom is not freedom to do what is irrational, or stupid, or wrong."[121] But in courtrooms, liberty rights mean freedom to do the irrational, stupid, even the wrong. That is why judges routinely honor nonrational, even irrational, choices that may cut against a decisionmaker's best interests.[122] Self-determination may even trump human life.[123] Judges accept the nonrational determination of Jehovah's Witnesses to die rather than accept blood transfusions. The mentally ill are not usually confined against their wishes unless they pose a threat to themselves or to others.[124]

Judges who deny personhood to every nonhuman animal act arbitrarily. They don't *say* they do. Instead, they use legal fictions. Legal fictions are transparent lies they insist we believe. These allow them to attribute personhood not only to humans lacking consciousness and even brains but to ships, trusts, corporations, even religious idols.[125] They pretend these entities enjoy autonomy. Legal scholar John Chipman Gray could not see any difference between pretending that will-less humans have autonomy and doing the same for nonhuman animals.[126] Because legal fictions may cloak abuses of judicial power, Jeremy Bentham characterized them as a "*syphilis* . . . [that] carries into every part of the system the principle of rottenness."[127]

A fair and rational alternative exists and it is this: most moral and legal philosophers, and just about every common-law judge, recognize that less complex autonomies exist and that a being can be autonomous if she has preferences and the ability to act to satisfy them. Or if she can cope with changed circumstances. Or if she can make choices, even if she can't evaluate their merits very well. Or if she has desires and beliefs and can make at least some sound and appropriate inferences from them.[128]

In *Rattling the Cage,* I called these lesser autonomies "realistic." I now think "practical" better describes them.[129] "Practical autonomy" is not just what most humans have but what most judges think is *sufficient* for basic liberty rights, and it boils down to this: a being has practical autonomy and is entitled to personhood and basic liberty rights if she:

1. can desire;
2. can intentionally try to fulfill her desires; and
3. possesses a sense of self sufficiency to allow her to understand, even dimly, that it is she who wants something and it is she who is trying to get it.

Consciousness, not necessarily self-conscious, and sentience are implicit in practical autonomy.

Human newborns, fetuses, even ovums, sometimes have legal rights. This might be explained as resulting from legal fictions or sheer arbitrariness. But it might have something to do with autonomy. They may not have it now, but it's believed they have the potential. And if they have the potential, the viewpoint holds that we should treat them as if they had autonomy now.

But the potential for autonomy no more justifies treating one as if one had autonomy any more, and probably less, than does one's potential for dying justify that one should be treated as if one were dead.[130] Philosopher Joel Feinberg thinks allocating rights based on potential is a logical error. Potential autonomy gives rise to potential rights. Actual autonomy gives rise to actual rights.[131] The potentiality argument moreover fails to explain how common law can grant dignity-rights to adult humans who never enjoyed autonomy, and never will.

Isaiah Berlin wrote, "If the essence of men is that they are autonomous beings . . . then nothing is worse than to treat them as if they were not autonomous, but natural objects, played on by causal influences at the mercy of external stimuli."[132] The same is true for any being who meets the requirements for practical autonomy. She is entitled to liberty rights. Because much of the world, certainly the West, links basic liberty rights to autonomy, and because autonomy is often seen as the foundation of human dignity, in *Rattling the Cage,* I called basic liberty rights "dignity-rights."[133] Keep these three sufficient conditions in mind as you read the discussions of the cognitive abilities of honeybees, dogs, dolphins, elephants, whales, orangutans, and gorillas in the coming chapters. Later, I will explore whether an animal may be entitled to equality rights even if she lacks practical autonomy.

Isn't Sentience Enough?

I have been criticized for arguing in *Rattling the Cage* that practical autonomy, not just the ability to suffer, entitles one to dignity-rights. One animal protection lawyer wrote that "[i]f one accepts the philosophies of Jeremy Bentham and Peter Singer, then an animal's ability to feel pain and suffer, and not its ability to count or use tools, should be the measuring point in extending legal personhood."[134] Law professor Cass Sunstein wrote in the *New York Times Book Review* that he was unsure why I spent so much space making the scientifically controversial argument that chimpanzees and bonobos are autonomous: "[W]ould cruelty toward [them] be justified if it turned out that (as some scientists contend) chimpanzees do not really understand American Sign Language? *Why isn't the capacity to suffer* a sufficient ground for legal rights of some kind—for dogs, cats, horses, chimpanzees, bonobos, or for that matter cognitively impaired human beings?"[135] A subsequent letter to the editor of the *Book Review* damned my arguments as "morally flawed" and insisted that "suffering, not intelligence, must be the only consideration"; hadn't Bentham said, "'The question is not, Can they *reason*? nor, Can they *talk*? but Can they *suffer*?'"[136]

If I were Chief Justice of the Universe, I might make the simpler capacity to suffer, rather than practical autonomy, sufficient for personhood and dignity-rights. For why should even a nonautonomous being be forced to suffer? But the capacity to suffer appears irrelevant to common-law judges in their consideration of who is entitled to basic rights. What is at least *sufficient* is practical autonomy. This may be anathema to disciples of Bentham and Singer. I may not like it much myself. But philosophers argue moral rights; judges decide legal rights. And so I present a legal, and not a philosophical, argument for the dignity-rights of nonhuman animals.

A New Deal for Nonhuman Animals

As we discuss the entitlement of nonhuman animals to dignity-rights, it is important to remember that the united voices of Aristotle and the Stoics were just one of many—Stoic, Aristotelian, Epicurian, Platonist, Pythagorean, Cynic, and others—competing in the Ancient World about animal minds and the place of nonhumans in the universe. But Western Christianity and law listened to just this one voice. With few exceptions, there are no more human slaves. Women and children are not legal things. But every nonhuman animal remains a thing, as she was 2,000 years ago. We will see that twenty-first century science tells us that the old way of seeing nonhuman animals as mindless was terribly wrong. It is time for animal law to enter the present.

THREE

Who Gets Liberty Rights?

A Scale of Practical Autonomy

Donald Griffin is the father of the scientific discipline of "cognitive ethology," which investigates and compares mental phenomena among animals. He knows how elusive firm conclusions about mental abilities can be and finds it helpful to think in probabilities: what are the chances an animal feels or wants or acts intentionally or thinks or knows or has a self?[1] The more certain we are that the answer to any of these questions is "yes," the closer the probability is to 1.0. If "no" is certain, the probability is 0.0. If we think the answer impossible to know, or that it's possible but we just don't know anything, the probability is exactly 0.50.

Griffin gives Kanzi, the bonobo whom I discussed at length in *Rattling the Cage,* as an example of a nonhuman animal for whom it would be reasonable to estimate the probability that he's conscious and possesses sophisticated mental abilities close to 1.0. Griffin concedes, upon present evidence, the probability that some of the million species of animals are simply not conscious (0.0). For many of the millions of intermediate cases, we can make only rough estimates. He thinks that often the best we may be able to do at present is to assign values below 0.50, exactly 0.50, above 0.50, and almost 1.0.[2]

I think the evidence permits us to assign a more precise "autonomy value" to each animal whom we discuss: Alex the Grey parrot, Christopher my son, Marbury his dog, Echo an African elephant, the dolphins Phoenix and Ake, Chantek an orangutan, and Koko the gorilla.

The more exactly the behavior of any nonhuman resembles ours and the taxonomically closer she is, the more confident we can be that she possesses desires, intentions, and a sense of self resembling ours, and that we can fairly assign her an autonomy value closer to ours. Nevertheless, practical autonomy is hard to quantify; in determining whether it exists and to what degree, and assuming that at least some association between general mental complexity and practical autonomy exists, we will consider not just mental abilities that speak directly to autonomy but mental complexity in general.

When a range of behaviors in an evolutionary cousin closely resembles ours, as they do with Kanzi, we can confidently assign a value of almost 1.0. We'll call Kanzi a 0.98 and place any animal having an autonomy value of .90 or greater into Category One. These animals clearly possess sufficient practical autonomy to qualify them for basic liberty rights. They are probably self-conscious and pass the mirror self-recognition (MSR) test. The standard MSR was developed by psychologist Gordon Gallup, Jr., in the 1970s when he worked with chimpanzees. Gallup placed red marks on the heads of anesthetized chimpanzees, then watched to see whether they touched the marks when peering into a mirror. He assumed that if they did, they were self-aware.[3] Gallup's MSR test and variants adapted for other nonhuman species and human infants are widely used as markers for visual self-recognition, though there is disagreement about what they signify and whether failures mean self-awareness is lacking. But if the subjects pass, they should be placed into Category One. These animals may also have some of, or all, the elements required for a theory of mind (they know what others see or know); this means that they understand symbols, use a sophisticated language or language-like communication system, and may deceive, pretend, imitate, and solve complex problems.

Into Category Two we'll place animals who fail MSR tests and may lack self-consciousness and every element of a theory of mind. These animals may possess a simpler consciousness, which means they can mentally represent and act insightfully; think; perhaps use a simple communication system; have a primitive, but sufficient, sense of self; and are at least modestly close to humans on an evolutionary scale. Insight is sometimes used as a synonym for thinking, for it allows a being to solve a problem efficiently and safely by mentally "seeing" the solution without having to engage in extensive trial and error.[4] Biologist Bernd Heinrich speculates that consciousness, which he believes implies awareness through mental visualization, developed "for one specific reason only: To make choices."[5] If he is right, insight is strong evidence of practical autonomy.

The strength of each animal's liberty rights claim will turn upon what mental abilities she has and how certain we are she has them. Category Two covers the immense cognitive ground of every animal with an autonomy value between 0.51 and 0.89. Whether an animal should be placed in the higher, middle, or lower reaches of Category Two depends upon whether she uses symbols, conceptualizes (mentally represents), or demonstrates other sophisticated mental abilities. Her taxonomic class (mammal, bird, reptile, amphibian, fish, insect) and the nearness of her evolutionary relationship to humans (which are related) may also be important factors.

We will assign taxonomically and evolutionarily remote animals, whose behavior scarcely resembles ours and who may lack all consciousness and be nothing but living stimulus-response machines, an autonomy value below 0.5. The lower the value, the more certain we can be they utterly lack practical autonomy, though they may, or may not, be eligible for equality rights. We'll place them all into Category Four. Finally, we do not know enough about many, perhaps most, nonhuman animals reasonably to assign any value above or below 0.5. Perhaps we have never taken the time to learn about them or our minds are not sufficiently penetrating to understand them. We will assign them to Category Three.

Consciousness is the bedrock of practical autonomy. We'll see in Chapter 4 that prominent neuroscientists believe emotions produce consciousness. If that is true, when did emotions and consciousness arise? Neuroscientist Antonio Damasio writes that the structures responsible for simple consciousness are "of old evolutionary vintage, they are present in numerous nonhuman animals, and they mature early in individual human development."[6] Ethologist Marian Stamp Dawkins argues that mammals and birds are conscious.[7] Nobel prize–winner Gerald Edelman thinks brain structures generating simple consciousness can be found in most mammals, in some birds, and perhaps in some reptiles; consciousness could be about 300 million years old.[8] Because emotional stress raises the body-core temperature and heart rate of rats, humans, birds, and reptiles, but not amphibians, physiologist Michel Cabanac argues that emotion and consciousness emerged between the evolution of amphibians and reptiles, more than 200 million years ago.[9] Reptiles, mammals, and birds are probably conscious; others, probably not. Donald Griffin finds any firm dividing line premature.[10]

A Category One animal clearly has practical autonomy and is entitled to the basic liberty rights of bodily integrity and bodily liberty. A Category Four animal probably, and perhaps clearly, lacks practical autonomy and is not entitled to liberty rights. A Category Three animal is there precisely because we know so little about her mental abilities that we can't make a rational judgment. The best we can do is recognize our present ignorance and limitations, determine to learn what we can, be alert to the evolving knowledge that will allow us to place these animals into another category, and take to heart what philosopher Martha Nussbaum has written: we may have to choose between generosity and mean-spiritedness.[11]

The Precautionary Principle

How should we assign an autonomy value to an animal whose mental abilities we know something about, but are uncertain? Scientific uncertainty exists when data are incomplete or absent because we can't or don't know how to measure accurately; or we sample improperly because our theoretical models are just wrong; or because we mistake something for cause and effect.[12] Scientists recognize that absolute scientific truth doesn't exist. Much of what scientists do is to try to gain more certainty.

Uncertainty is no less common in law. But courts lack the luxury of deferring judgment until data are more complete. In the face of uncertainty and chance of error, judges must decide and content themselves with deciding on which side they wish to err. In Anglo-American law a criminal defendant is presumed to be innocent until a jury of twelve unanimously finds him guilty of committing the crime charged beyond a reasonable doubt. Every reasonable doubt is resolved in favor of the defendant. On the other hand, the American Fugitive Slave Act of 1850 did not allow an accused fugitive slave the usual trial by jury or the chance to testify. A summary hearing, often thirty minutes, sometimes five, was held before a specially appointed commissioner instead of a regular judge; the decision was unappealable, and no court could issue a writ of *habeus corpus* or delay the trip south.

Tort law normally permits an action unless it can be shown to cause harm.[13] Just as the Fugitive Slave Act Law assumed an accused fugitive slave really was a fugitive, an "assimilation principle" assumed the environment could incorporate vast insults, pollutants, poisons, and toxics without serious effect.[14] Scientists know that this assumption is untrue. Policymakers slowly responded by adopting a "precautionary principle." It constituted a "fundamental shift" in how environmental concerns were faced.[15] The precautionary principle rejects science "as the absolute guide for the environmental policy maker" and "embodies the notion that where there is uncertainty regarding the potential impact of a substance or activity, 'rather than await certainty, regulators should act in anticipation of possible environmental harm to ensure that this harm does not occur.' Without such an approach, an activity or substance might have an irreversible impact on the environment while scientists work to determine its precise effect."[16] At the principle's core is the idea that nonnegligible environmental risks should be prevented, even in the face of scientific uncertainty.

The precautionary principle is finding a home in United States and German law; it is emerging in English, Australian, and European law and evolving as a customary rule of international environmental law.[17] For example, the United States Endangered Species Act requires federal agencies to give the benefit of the doubt to threatened or endangered species when determining how to act, while the Marine Mammal Protection Act permits takings of marine mammals "only when it is known that the taking would not be to the disadvantage of the species."[18] The World Charter for Nature says "activities which are likely to pose a significant risk to nature shall be preceded by an exhaustive examination; their proponents shall demonstrate that expected benefits outweigh potential damage to nature, and where potential adverse effects are not fully understood, the activities should not proceed."[19]

For centuries, law followed an "exploitation principle" in its dealings with nonhuman animals. All nonhuman animals were erroneously thought to lack every sophisticated basic mental ability—desire, intentionality, self, probably even consciousness—and were categorized as legal things, mercilessly exploited. But evidence is clear that at least some do not lack basic mental abilities. In light of what we know, it is time to apply a precautionary principle to the law of nonhuman animals. Depriving any being possessed with practical autonomy of basic liberty rights is the most terrible injustice imaginable. When there is doubt and serious damage is threatened, we should err on the cautious side when some evidence of practical autonomy exists. And some evidence is required, for every version of the precautionary principle instructs "how to respond when there is some evidence, but not proof, that a human practice is damaging the environment."[20] Speculation is not enough.[21]

The precautionary principle has at least seven senses.[22] At its weakest, it requires that a decisionmaker merely think ahead and act cautiously. A stronger version requires a decisionmaker to regulate her actions carefully, even in the face of insufficient scientific evidence of a threat. Stronger still is the demand that the proponent of a potentially harmful act prove its harmlessness to an unusually stringent degree.[23] This shifting of the normal burden of proof may be tantamount to forbidding the act, for the more uncertain the evidence, the more likely it will be that whoever has the burden of proof will lose.[24] I have excellent firsthand experience.

In the 1990s, I used the American Endangered Species Act to try to stop a deer hunt at the once-pristine Quabbin Reservation in Massachusetts, unhunted for half a century. Deer are not endangered. I argued that shooting Quabbin deer with lead slugs might lead to the deaths of bald eagles, who could ingest the poisonous lead. Bald eagles were endangered. I proved that bald eagles often die from ingesting lead and that they could ingest the lead slugs lodged in the deer who would inevitably evade hunters and die where only bald eagles could find and eat them. But I could not prove the Quabbin eagles would be harmed by the lead in the dead Quabbin deer. Even if a Quabbin eagle dropped dead of lead poisoning in the middle of the park, I could not prove that the specific lead the eagle had ingested came from a hunted Quabbin deer. Maybe it came from a deer just across the way or the next state over.[25]

A kind of precautionary principle has been argued as a reason for not using seriously defective human beings in painful biomedical research. The reason, philosopher Christina Hoff wrote, is not because they are human, for Hoff concedes that is insufficient. It is because we cannot "safely permit anyone to decide which human beings fall short of worthiness. Judgments of this kind and the creation of institutions for making them are fraught with danger and open to grave abuse."[26] When rights for nonhuman animals are involved, there is a compelling reason to apply the precautionary principle, and it goes even beyond Hoff's reasoning. However, it doesn't yet exist in environmental law.

At the end of the seventeenth century, John Newton, a former London slaving captain, wrote, "I can't think there is any intrinsic value in one colour more than the other, that white is better than black, only we think it so, because we are so, and are prone to judge favourably in our own case."[27] We should have little confidence in the fairness of a decision reached by a judge having a personal stake in the outcome. This human proneness "to judge favourably in our own case" explains why one of the oldest maxims of the common law is no one may judge her own case.[28] In the seventeenth century, Lord Coke fixed this into the common law when he voided the right of the Royal College of Physicians to fine medical practitioners, because the college retained half of every fine levied.[29] The United States Supreme Court prohibits a judge from having even the "slightest pecuniary interest" in the subject matter of his decision.[30] Whose blood doesn't rise when reading that in the years before the Civil War, the slaveocracy-dominated Congress of the United States enacted a Fugitive Slave Act that paid the deciding magistrate twice as much for flinging the poor wretch back into bondage than he received for declaring her free?[31]

If judicial attitudes reflect society's, most judges believe their health or the health of their families and friends may depend, in part, upon the use of nonhuman animals in biomedical research. They probably eat animal flesh, drink animal fluids, wear animal skins and fur, take their children to circuses to enjoy the "performances" of captive nonhuman animals, hunt them, and participate in some of the numerous ways in which society exploits nonhuman animals. Judges have a personal stake in the outcome of cases that decide whether nonhuman animals have legal rights in much the same way that judges of the antebellum American South had personal stakes in the outcome of slavery cases. They will inevitably be biased, and their biases will just as inevitably infect their decisions.[32]

Such conundrums are typically solved by invoking the "Rule of Necessity," which states that if all judges are disqualified from deciding a case, none are.[33] But that does not give judges license to indulge their biases. To the contrary. Formal use of the precautionary principle is necessary just to counteract judicial bias. Judges ruling from necessity must, to rule as fairly as they can, exert every ounce of moral strength, every particle of objectivity they possess, always keeping in mind that they are prone to decide in their own favor and that long-standing inequities have, in law professor Laurence Tribe's words, "survived this long because they have become so ingrained in our modes of thought; the U.S. Supreme Court recognized a century ago that 'habitual' discriminations are the hardest to eradicate."[34]

Assigning Autonomy Values

Because it appears likely that many, perhaps most, mammals and birds have emotions, are conscious, and have selves, the burden of proving at trial that an individual mammal or bird lacks practical autonomy should be shouldered by the one who wants to harm them. On the other hand, when someone objects to the proposed harm of an animal neither mammal nor bird, the burden of proving practical autonomy should rest with him.[35]

After reviewing the evidence and taking into account who has the burden of proof, a judge could assign an autonomy value to the nonhuman animal. Unbiased judges may honestly differ, especially as data become less certain. How certain is she that an animal possesses practical autonomy? How strong a scientific argument has been made? How valid are the data? Have they been replicated? Have data from multiple areas of inquiry begun to converge? How cautious does this judge want to be? Remember that a value of 0.9 or higher will place an animal in Category One; he will be presumed to have practical autonomy sufficient for basic liberty rights. A Category Four animal, scored less than 0.5, will be presumed to lack practical autonomy. Any animal given a score of 0.5 means we haven't a clue.

There may be vast differences among Category Two animals, whose autonomy values range from 0.51 to 0.89. A refusal to apply the precautionary principle means that every Category Two animal would be disqualified from liberty rights. An expansive application of the precautionary principle means that an animal having an autonomy value above 0.50 should be granted rights. I propose an intermediate reading: any animal with an autonomy value higher than 0.70 is presumed to possess practical autonomy sufficient for basic liberty rights. But how should we treat animals who score higher than 0.50 but less than 0.70? By definition, some evidence exists that they possess practical autonomy; but it is weak, either because at least one element is missing, or the elements together are feebly supported.

More than a million animal species exist. Darwinian evolution postulates a natural continuum of mental abilities, but the animal kingdom is incredibly diverse. At some taxonomic point, the elements of practical autonomy begin to evaporate: self, intentions, desire, sentience, and finally consciousness. We don't know precisely where. I have stood on the summit of Cadillac Mountain on Mt. Desert Island in Maine watching the summer sun rise. At four o'clock in the morning, it was indisputably night; at seven o'clock, indisputably day. When did night become day?

We could deal with this problem in one of two ways. Not using the precautionary principle would allow a judge to draw a line and an animal beneath it would not be entitled to liberty rights. But another way is consistent with even a moderate reading of the precautionary principle. Personhood and basic liberty rights should be given in proportion to the degree one has practical autonomy. If you have it, you acquire rights in full; if you don't, the degree to which you *approach* autonomy might make you eligible to receive some proportion of liberty rights.[36]

This idea of granting proportional liberty rights accords with how judges often think. They may give *fewer* legal rights to a human who lacks autonomy, but they don't make her a thing. A severely retarded human adult or child who lacks the mental wherewithal to participate in the political process may still move about freely. Judges may give *narrower* legal rights to her. A severely mentally limited human adult or child might not have the right to move in the world at large, but may move freely within her home or within an institution. Judges may give *parts* of a complex right (remember, what we normally think of as a legal right is actually a bundle of them). A profoundly retarded human might have a claim to bodily integrity but lack the power to waive it, thus being unable to consent to a risky medical procedure or the withdrawal of life-saving medical treatment.[37]

Consistent with a moderate use of the precautionary principle, we need not grant basic liberty rights to a nonhuman animal who has just a shadow of practical autonomy. But it would be consistent with such a reading for an animal with an autonomy value of 0.65, perhaps even 0.60, to be given strong consideration for some proportional basic liberty rights.

At some point, autonomy completely winks out and with it nonarbitrary entitlement to liberty rights on the ground of possessing practical autonomy. Although judges and legislators might decide to grant even a completely nonautonomous being basic liberty rights, it's hard to think of grounds upon which they might do it nonarbitrarily. We will see in Chapter 12 that this strengthens the argument that as a matter of equality, nonautonomous animals of many species should be entitled to basic rights, too.

Human Yardsticks for Nonhuman Animals

Nature writer Henry Beston wrote "the animal shall not be measured by man. In a world older and more complete than ours they move finished and complete, gifted by extensions of the senses we have lost or never attained, living by voices we shall never hear."[38] However, the liberty rights to bodily integrity and bodily liberty are embedded in law precisely because they are basic to human wellbeing, and the autonomy values we assign to nonhumans will be based upon human abilities and human values. This argument bridges the present, when law values only human-like abilities but must still confront the argument for the liberty rights of nonhuman animals who have them, and the future, when the law may value nonhuman abilities as well. For the present, I accept that the law measures nonhuman animals with a human yardstick.

But just because law is so parochial, we mustn't think human intelligence the only intelligence. Intelligence is a complicated concept that intimately relates to an ability to solve problems.[39] Biologist Bernd

Heinrich says "we can't credibly claim that one species is more intelligent than another unless we specify intelligent with respect to what, since each animal lives in a different world of its own sensory inputs and decoding mechanisms of those inputs."[40] Dolphin expert Diana Reiss argues that intelligence cannot properly be conceived solely in human terms and condemns any assumption that "only our kind of intelligence is 'real intelligence.'"[41] We mustn't think human self the only self or human abilities the only important mental abilities.

In a famous essay, "What is it like to be a bat?" philosopher Thomas Nagel focused on bats, instead of flounders or wasps, because they are mammals, and "if one travels too far down the phylogenetic tree people gradually shed their faith that there is experience at all."[42] Even bats, Nagel wrote, are "a fundamentally *alien* form of life."[43] How can we know when a nonhuman possesses practical autonomy? In *Rattling the Cage,* I answered with "de Waal's Rule of Thumb": in the absence of strong contrary arguments, if closely related species act in the same way, it is likely, though not certain, they share the same mental processes.[44] Because chimpanzees and bonobos are evolutionary first cousins and have senses identical to ours, we can feel confident that our basic interests are similar. But the likelihood of misunderstanding or mistaking basic interests increases with evolutionary distance and differences in ecology, information-processing capabilities, and the ways in which we perceive or conceive of the world. Humans, for example, can live happy solitary lives. But a lone Atlantic bottle-nosed dolphin is scarcely a dolphin. Humans don't need to swim. But even after seventy generations of being factory-farmed in cages, minks expend large amounts of energy just for the opportunity to swim, and when they can't, their bodies produce high levels of cortisol, a "stress" hormone, similar to that produced when they are denied food.[45]

Elephant researcher Joyce Poole asks us to imagine showing drawings of human and elephant heads to an extraterrestrial. It would be reasonable to assume the creature with the huge ears and nose has terrific hearing and smelling abilities, while the one with the tiny nose and big eyes can see, but is unlikely to smell or hear as well. This would be correct. Elephants hear sounds below the human threshold.[46] Elephants' nasal cavities contain seven of the turbinals specialized for odor detection. Poole concludes that elephants, who also detect hormones with a unique Jacobsen's organ, are so different from us that they live in another sensory world.[47] More than forty years ago, psychologist Harry Harlow, reviled today by many for his cruel deprivation experiments on baby monkeys, wrote "it is hazardous to compare learning ability among animals independently of their sensory and motor capacities and limitations."[48] Harry Jerison says "the nature of self is surely significantly different in different species," but that human intuitions about self "are so strong that it is difficult to imagine a creature with information processing capacities comparable to ours, equal to us in intelligence . . . that has a differently constructed self." But they exist and he gives dolphins as an example.[49]

The animals who fall into Categories Three and Four, and some in Category Two, may not now be entitled to basic liberty rights simply because we don't value their brand of intelligence, their style of learning, their sense of self. But Category One animals, and at least some of those in Category Two, measure up even to human standards. Chimpanzees and bonobos certainly do. In the next seven chapters, I will act the judge, review the evidence, and make my own judgments about whether practical autonomy exists, how strong a scientific argument has been made, and how valid the data are. I will apply the spirit and letter of the precautionary principle to uncertainties, then assign an autonomy value based on that evidence to honeybees, my son, Christopher, Christopher's dog, Marbury, Echo the African elephant, the dolphins Phoenix and Ake, Chantek, an orangutan, and Koko the gorilla.

FOUR

Christopher

In Flux

At Berkeley's Cafe Venezia one chilly November evening in 2000, I delighted in a plate of gnocchi smothered with California porcini mushroom sauce. But my three dining companions, Jonas Langer, Sue Taylor Parker, and Anne Russon, delighted me even more. Together they had vast experience contemplating the complexities of children's minds. As much as we know about chimpanzee and bonobo minds, it's nothing compared to what we know about how the mind of my four-year-old son, Christopher, developed. When did he first desire? When did he begin to act intentionally? When did he gain a self? How do we know?[1]

Jonas Langer, a child developmental psychologist at Berkeley, has an interest in nonhuman primate cognition. Anne Russon is a comparative developmental psychologist from Toronto who studies rehabilitant orangutans in Borneo. Anthropologist Sue Taylor Parker from Sonoma State explores cognition in great apes. Langer and Parker are devotees of the world's most famous, and controversial, child development psychologist, Jean Piaget. I asked them to dinner because child development theory is in flux. Biological-maturation theorists think development springs from within, moving in genetically programmed ways, perhaps with environmental triggers. Environmental-learning theories find environment the most important, learning critical, with development occurring more or less seamlessly. Constructivist thinkers, such as Piaget, believe that nature and nurture play equal roles, with children constructing higher levels of knowledge as they actively master their environments. Piaget also believed that development occurs in stages. A fourth, related theory is that nature and nurture don't interact directly, as they do in the other three theories, but indirectly, through culture.[2] Michael and Sheila Cole, authors of a recent text on child development, conclude that no single theory fits the facts. Instead, leading theories "might best be thought of as filters with each theory highlighting certain features of the overall development process."[3]

Scientists clash over which abilities are innate and which are learned. Are minds divided into discrete units, constructed of modules each with a different job, that click away from birth? Do modules exist only in humans?[4] Can they exist in other animals? Are some modules furnished with a delayed action fuse that causes them to lie dormant for months, even years, then explode onto the mental scene? Does experience organize minds into modules?[5] Or are minds not modular at all, perhaps equipped with "core knowledge" about the world?

Today, no scientist thinks Christopher's mental abilities are entirely learned or completely innate. Few believe that more than a handful are innate; fewer still think that biological foundations for such a complex cognitive ability as language exist or are either innate or restricted to human beings.[6] Most concede that some innate abilities exist, the most important being a powerful learning mechanism. Whatever Christopher's innate abilities, they probably ripened through experience hammered upon the anvil of this learning mechanism.[7] But no one agrees how it works.[8] Elizabeth Spelke, a Harvard developmental psychology professor, nicely sums the disarray: "There is no consensus . . . about when knowledge begins, what it consists of, how it manifests itself, what causes it to emerge, how it changes with growth and experience, or what roles it plays in the development of thought and action."[9] Don't expect consensus about nonhuman minds.

In an early draft of this chapter, I described Christopher's development primarily in Piagetian terms. One psychologist has written that "[a]ssessing the impact of Piaget on developmental psychology is like assessing the impact of Shakespeare on English literature or Aristotle on philosophy—impossible. The impact is too monumental to embrace and at the same time too omnipresent to detect."[10] Modern developmental psychologists instinctively measure their work against Piaget's, though many scientists, including biological anthropologist Annette Karmiloff-Smith, believe that Piaget's entire theory of developmental structure was misconceived.[11] That's why in the fall of 2000 I sent the draft to Jean Mandler, a prominent biological-maturation theorist, because I knew she would disagree with Piaget and I wanted to understand why. Though intense autumn rain and hail had damaged the roof and walls of her Hampstead Hill home in London, she obliged.

She didn't know, she said, "any major infant psychologist in the United States or England who believes Piaget's stage theory of infancy, or many of its details either."[12] Piagetian theory "is dying but it is not entirely moribund."[13] With permission, I passed Mandler's detailed criticisms to six prominent scientists who study nonhuman animal cognition. Three agreed, three didn't. A few weeks later, I was munching on gnocchi and listening to Langer and Parker defend Piaget.

Langer is a thin balding man in his sixties with a ready sense of humor and an accent lightly reflective of his native Belgium. He has little truck with environmental-learning thinkers. Behaviorism, he said, was in vogue when he was being trained. Stimuli and responses were measured. Consciousness was irrelevant. So were minds. Everything was learned. Biological-maturation thinkers, perhaps influenced by the thinking of MIT's Noam Chomsky, theorized that many of a child's basic intellectual abilities are present at birth, or soon after. Langer said they think almost everything is mental, an overreaction to the behaviorism that saw nothing as mental. With a wolfish grin, he asked how they can call themselves child *development* psychologists. They don't think minds progress in stages, he says, because they hardly believe in development at all! Langer made two fists. Because Piaget steered a middle course between dominant nurture and dominant nature, he gets blasted from both sides. Langer crashed his fists over his pasta.

Psychologist Henry Wellman, who disagrees with many Piagetian conclusions, still thinks he is often targeted precisely because his work is so persuasive.[14] Even Piaget's critics concede he's often misunderstood.[15] And it's not just child development scientists who look to his work; nonhuman animal cognition researchers often find Piaget, with all his limitations, helpful.[16] That is why the two most recent and comprehensive reviews of primate development are shot through with Piaget.[17] Biologist Kim Bard writes, "Piaget's theory of cognitive development in humans from birth to adulthood remains one of the core theoretical perspectives in developmental psychology. Although many may disagree with particular aspects of the theory (e.g., development is not stagelike) or the database (i.e., the ages assigned by Piaget are too old), the behavior patterns of the sensorimotor period (birth to 2 years) are one of the areas where there is very little that is controversial."[18]

Sue Taylor Parker goes further: Piaget's "is probably the *only* suitable . . . framework" for comparing human and nonhuman primates.[19] Anne Russon agrees, at least for now, because Piaget's is the only model used in both human and nonhuman studies.[20] Irene Pepperberg thinks Piaget can help us make valid comparisons not just among primates but between them, nonprimate mammals, and the African Grey parrots she has studied for more than twenty years.[21] For these reasons, I situate Christopher within a modified Piagetian framework.

Christopher's Mind and How It Grew (According to Piaget)

Piaget postulated four stages of cognitive development. The "sensorimotor period," with six substages (really convenient discussion points), begins at birth and ends at about age two. Six basic skills (object, space, time, causality, imitation, and circular reactions) are acquired in lockstep, seamlessly, without pause, and in the same order.[22] We will focus on the sensorimotor because the vast majority of nonhuman animals fall within it.

According to Piaget, "preoperations" signaled the onset of Christopher's reflective life (Mandler: "The notion of a preoperational stage was discarded 15–20 years ago").[23] These were his first steps on the journey from *being* his perceptions to *having* them, of using insight to solve problems.[24] As I mentioned, insight is often synonymous with thinking, for it makes it possible to solve a problem mentally, efficiently, and safely by "seeing" the solution without engaging in extensive trial and error.[25] This stage has two subperiods. The "symbolic period" began about the time he turned two. It will lead to his "intuitive subperiod" near the age of four.[26] (Mandler: "There is no such thing as an intuitive subperiod").[27] Christopher thinks that objects appear to him as they are and everyone understands them as he does.[28] His thought, increasingly laden with images and symbols, remains relatively concrete and static, inflexible and slow.[29] But his symbols are increasing in complexity as immature concepts emerge.[30]

When he is about six or seven, Christopher will enter the "concrete operations" period. The child psychologist John Flavell says that Christopher will give "the decided impression of possessing a solid cognitive bedrock, something flexible and plastic and yet consistent and enduring, with which he can structure the present in terms of the past without undue strain and dislocation, that is, without the ever-present tendency to stumble into the perplexity and contradiction which mark the pre-schooler."[31] Problems that used to stump him may now appear simple. Later, his "formal operations" period will be marked by complex mental abstractions. By the age of twelve, he will be almost freed from perception's shackles, able to concentrate upon the what-might-be, rather than the what-is.[32]

You need not agree with Piaget to sense that young Christopher inhabits a universe somewhat different from yours.[33] But his is no "random fanc[y], incomplete or dim perception of reality as we see it." It is "a distinct separate reality, with a logic, a consistency, an integrity all its own."[34] The same may hold true for many nonhuman animals.

The Perceptual Life: Intelligence in Action

No formal operations period abilities have been detected in a nonhuman. Chimpanzees who understand something about the conservation of liquid volume or solid quantities may spike into the concrete operations stage. Most animals never advance beyond the sensorimotor. But they achieve what Piaget called "practical intelligence."[35]

In the sensorimotor stage, Christopher combined his senses with such developing motor skills as sucking, reaching, and grasping, to perceive, then act upon, objects he perceived. But Piaget thought his mind could not symbolize and he could not consciously think about what he was not directly perceiving. Instead of "concepts," he had "percepts."[36] Because many, if not most, nonhuman animals may have "percepts" only, we must understand what practical intelligence is, and what this is not.

Sensorimotor intelligence is "intelligence in action," an "unreflective, practical, perceiving-and-doing sort of intelligence."[37] The universe of a sensorimotor animal is perception;[38] he can experience everything a moment packs.[39] This does not mean he can't recall the past or imagine a future. He just can't do it unchained from the present. The sight, the sound, the smell, the taste, or the feel of an object can trigger memories of what was and some sense of what is to come.[40] Once he can imagine things, he is not then perceiving; he has concepts.[41] Conceptual intelligence strives for truth, while sensorimotor animals achieve concrete goals, such as picking up a ball.[42] One passes from perceptual "actions *in* reality" to conceptual "representations *of* reality."[43] Sensorimotor intelligence is the difference between "knowing" and "knowing how." I took piano lessons as a child, but I refused to practice and did not keep up my playing. Today I can sit at a keyboard and play snatches of songs I learned years ago. But I can't tell you how I do it. I don't know any melodies. I sit at the keyboard and let my fingers go until they outrun their memory.[44]

Mandler says because "sensorimotor information is not accessible does not mean that you have no awareness of the sensations . . . involved in perceptual and motor learning." It "should not be taken as indicating a zombie-like lack of awareness. Rather it means that you have not *conceptualized* the sensorimotor information and so cannot think about it. You see a tree is green, you experience a greenness, but that is not the same as thinking 'This tree is green' or 'Trees are green.'"[45] According to Piaget, she says, the sensorimotor is a life "not thought, only lived. Sensorimotor schemas enable a child to walk a straight line but not think about the line in its absence, to recognize his or her mother but not to think about her when she is gone."[46] Says psychologist Robert Kegan, "It is not that the child [i]s not a meaning-maker or that it [i]s unable to think. Indeed it would even be fair to say that the child did its thinking by moving and sensing, that the body *was* its mind."[47] As a child begins to *have* reflexes, sensations, and movements "rather than be them, he stops thinking he causes the world to go dark when he closes his eyes."[48]

Some psychologists argue that concepts and percepts fall on a continuum, concepts being percepts swollen with information.[49] More believe they differ in kind, though they dispute when and how concepts appear. If there is an exclusively sensorimotor period, Mandler says, it can't last more than a few months. She thinks the shift to concepts begins no later than the age of three or four months and in parallel to, but separate from, the sensorimotor system.[50] Depending upon which theory you accept, percepts are more like concepts than most development psychologists think; babies develop percepts and concepts at the same time, as Mandler thinks; or the sensorimotor period is much shorter than Piaget imagined. But everyone agrees Christopher began to want things, intentionally tried to get them, and developed a sense of self. The question is when.

Christopher Develops (According to Piaget)

Piaget said Christopher's six cognitive skills passed simultaneously through six sensorimotor stages, during which he began to perceive and move in increasingly more complicated ways, his memory becoming more critical to success. Substage 1, from which he emerged at about the age of one month, was saturated with reflexes: sucking, crying, swallowing; moving of eyes, arms, and legs.[51] Substage 2 lasted until about four months; during this time, he began to perfect his reflexes and coordinate and integrate his senses. He turned his head toward sounds and reached to touch objects. "Primary circular reactions," accidentally discovered, appeared. Merely actions that infants repeat, circular reactions are important in Piaget's infant development scheme, primary circular reactions being the simplest. Accidentally discovered by the baby, they are repeated not to attain a goal or change his environment, but for their own sake, such as sucking his thumb or mother's nipple. There was vague imitation.[52]

Christopher's universe was his extended self. Out of sight was out of existence.[53] He hadn't the faintest idea that when two things happened at about the same time, one might cause the other.[54] When I dropped his Barney the purple dinosaur doll on the floor, he looked to where I released it. His desires were amorphous and he probably thought he could make things happen by wishing them. If I did one of the few things he could manage, such as sucking my thumb, he might imitate me.[55]

Substage 3 ran from about four to eight months. Intentional actions stirred. Christopher might deliberately reach toward a desired goal, though he might not have been conscious that he desired it (there is a difference). Once, he banged his *Goodnight Moon* boardbook against the crib slats. Amused, he did it again, and again.[56] "Secondary circular reactions," Piaget's "procedures for making interesting sights last" emerged. He accidentally shook his rattle, then did it again and again.[57]

Christopher began to understand how objects relate to his body, though in a mechanical way.[58] When I dropped Barney, he leaned over to inspect the floor. When I covered Barney with his blanket, he forgot about it, even when he was holding it! If just Barney's head peeked out, he reached for the whole thing. But when I covered Barney entirely, he stopped.[59] He clearly imitated me, but only actions he knew or could see or hear himself perform, such as opening and closing his hand.[60]

Substage 3 bridged unintentional and intentional behavior where Christopher might stumble upon a means to an end and repeat it. But he was not yet acting intentionally. Intentional behavior required that he determine a goal, then try to achieve it.[61] Flavell calls what Christopher was doing "semi-intentional." He displayed a primitive "object permanence," realizing that objects exist, have substance, and maintain identities when out of sight.[62] This was the beginning of mental representations that would bloom at Substage 6, when he would be able not only to keep an absent object in mind but to reason about it.[63] His behavior suggested an early ability to classify, grouping books with books and blocks with blocks, and to seriate, placing smaller blocks atop larger.[64] He now understood that causes attained ends but acted as if only *his* causes attained them.[65] He sent trucks and balls sailing to see how they moved.[66] And he began to imitate behaviors he had never seen or he could not see, such as opening and closing his mouth.[67] He began to imitate words and seemed vaguely able to understand some.[68]

He entered Substage 4 at eight months and remained there until his first birthday. Now he began to exhibit true intentional behavior.[69] When I shoved his Barney doll under his blue blanket, he searched for it, sometimes ripping the blanket away. But his quests were restricted in a remarkable way. If I slowly and obviously hid the doll beneath his twin sister Siena's pink blanket, he rummaged for it beneath his blue one. When he couldn't find it, he gave up.

This "A-not-B" search error has been replicated hundreds of times. The error seems obvious. But the problem is harder than it appears. Finding the doll requires planning, inhibition, working memory, and some minimal attention span, all of which stretch eight-month-old human minds.[70] Gorillas, chimpanzees, monkeys, and African Grey parrots make this error; cats and dogs don't.[71] Piaget thought Christopher was not relying on what he saw, but repeating what had worked for him. His experience in finding Barney beneath the blue blanket overwhelmed his evolving notion that objects could exist apart from his experience of them, as he continued to think the only existing objects were those he was perceiving that instant.[72] Other scientists suspect "A-not-B" errors come from memory deficiencies or a half-dozen other possibilities.[73] Around his first birthday, Christopher stopped making these errors. But a different search restriction appeared. If I dropped a green Lego under his blanket and removed my hand, he searched for it only in my hand.[74] He knew that objects he could see could move about, but he didn't quite grasp that things he couldn't see could also move.[75]

Marbury's (the family dog) happiest day arrived when Christopher discovered the trial and error that characterizes Substage 5. He set off on what Piaget called "a search for novelty," through which he discovered "new means through active experimentation."[76] Cheerios snowed from his high chair as he slid, dropped, rolled, threw, pulled, and pushed them overboard . . . to see what would happen.[77] He dropped Barney, my wallet, and his mother's watch into Marbury's water dish . . . to see what would happen. He splashed cordless telephone receivers into the toilet . . . to see what would happen—until we accumulated a wicker basket stuffed with lifeless receivers.

Substage 5 spanned the first six months of Christopher's second year. He imitated novel facial and hand movements with greater fidelity.[78] Cause and effect sharpened, so he knew that by only touching Barney he would make the doll move; he was not just *wanting* to touch it. He realized that others could make things happen. If I put Barney on a blanket beyond his reach, he pulled the blanket toward him; and he would stick a toy hammer beneath a chair to scoop out a fallen plastic ball.[79]

But if I dropped a Lego under his blanket and removed my hand, he still searched for it in my hand, not under the blanket.[80] He couldn't grasp "invisible displacements."[81] Substage 5 was Christopher's last true sensorimotor stage, for Substage 6 would connect his sensorimotor and preoperations periods.

According to Piaget, Substage 6 arrived at about age eighteen months. Now Christopher could mentally represent unhitched from the moment, mulling solutions to problems through mental trial and error.[82] If I rolled a ball under the sofa, he figured out where it went and even detoured around obstacles to find it. When I stuck a Lego under his blanket, then Siena's, he searched my hand, then Siena's blanket, then his blanket, until he found it.[83] He determined a cause from seeing its effect.[84] Pretend-play blossomed. He not only imitated the movement of objects and people but engaged in the "deferred imitation" of actions such as imitating a face his mother made, but after she left the room.[85]

Piaget compared Christopher's early mental images to a slow-motion movie. He might understand each frame, but couldn't figure out the movie.[86] Cognitive scientists don't agree (of course) how to define "representation." Piaget reserved "representation" for Substage 6 kids.[87] But it seems safe to venture that representation means, in its broadest sense, one thing mentally stands for another. When we represent, we can think about ideas and things not before us; we can live in space and time where we are not, even if we can't use symbols.[88] Even developmental psychologists who disagree with Piaget's ideas about how Christopher attained his ability to represent, or think symbolically, agree that near the end of his second year, his thinking changed.[89]

Christopher Thinks

Fifteen years ago, psychologists developed three similar tests for infants: "expectancy-violation" tests, "habituation/dishabituation" tests, and "familiarization/preferential-looking" tests. These tests assume that infant familiarity breeds contempt, or at least boredom. Babies, they hypothesized, tend to look longer at interesting things and events. We can learn what a baby "expects" by measuring how long she looks. When something surprises her or fails to meet her expectations, she pays attention.[90]

Here is how one such test worked. A researcher showed six-to eight-month-olds a toy car that ran down an inclined track, disappeared behind a screen, and reappeared on the other side. She demonstrated this until the infants got good and bored. Then she raised the screen to reveal one of two situations. A large box either blocked the track or was just to one side. She lowered the screen and sent the car barreling down the incline to disappear behind the screen again. The babies stared longer when the car seemed to have gone through the box blocking the track.[91] In similar tests, infants of ten weeks appeared to expect that when one object is struck by another, it moves, and two objects can't be in the same place at the same time.

Unless pure sensorimotor representation allows for these findings—and Mandler, for one, doesn't see how—results that the car-and-track experiment produced seem to require Substage 6 representation.[92] If so, babies mentally represent by the time they are three or four months old.[93] In Mandler's view, Piaget's demand that infants reach for objects "may have produced a major misunderstanding about what infants know."[94] Before seven months, infants can't "demonstrate intentional behavior because of lack of ability to manipulate object effectively."[95]

Indeed, evidence is accumulating. Christopher represented and began thinking at a younger age than Piaget thought he could. Mandler has no doubt young babies represent and therefore think.[96] Three prominent child development psychologists don't even ask *whether* babies think, but do they "think two things are the same or different? And if they think two things are different, do they prefer one to the other?"[97] Other psychologists have marshaled impressive evidence to show that an object out of four-month-old Christopher's sight was not out of his Substage 3 mind. Seven-month-olds, ready to enter Substage 4, remember where an object was hidden ninety seconds before.[98] Infants engage in Substage 6 deferred imitation by fourteen months;[99] even if it isn't the sophisticated mental representations attained at Substage 6, mental residue lingers after an object disappears.[100] At Substages 4 and 5, Christopher seemed to understand that objects he wasn't perceiving existed. But his mental images were probably static, like photographs, and he couldn't imagine their movement.[101]

But no baby has confirmed what she "thinks." Piaget did not use expectancy-violation tests precisely because he thought they weren't stringent or subtle enough.[102] They allow a choice: baby looks or baby doesn't look. They give no clue *why* baby looks or doesn't look.[103] But when baby *reaches* for something, Piaget thought we can be more certain that she wants the object and thinks it exists at that spot.[104] To confuse matters further, researchers redoing some expectancy-violation tests have produced opposite results. Piagetians claim vindication, environmental-learning theorists argue the results support them, and biological-maturation theorists insist their theories have suffered only slight damage.[105] The Coles tersely conclude "there is no consensus on how much innate knowledge to attribute to small infants."[106]

I think the facts show Christopher had some innate tendencies and at birth knew more about how the world works than Piaget thought. At least some of his cognitive skills developed faster than Piaget thought they could, and not in lockstep.[107]

Christopher Desires and Acts Intentionally

Practical autonomy requires intentional agency. An intentional agent chooses her goals and the means to pursue them and monitors her progress toward them, persisting and changing strategies if necessary.[108] "Intentionality" is a muddy word. Philosophers use it to mean "aboutness": mental states refer to something.[109] That's why philosopher Daniel Dennett says that "thermostats, amoebas, plants, rats, bats, people, and chess-playing computers are all intentional systems." The key to your front door, the lock itself, have, says Dennett, a primitive intentionality.[110]

Many argue that true intentionality demands consciousness. Neuroscientist Susan Greenfield believes every conscious state is intentional.[111] Philosopher John Searle restricts true intentionality to most humans and some nonhuman animals.[112] Thermostats, locks and keys, amoebas, plants, and chess-playing computers only have an "as-if intentionality," behaving "as if" they have true intentionality. But they don't.[113] Truly intentional beings must be able to conjure up *some* mental representation of what they are being intentional "about," even if it's not a picture or word. Luckily, common understanding and philosophical understanding usually converge on about the same thing.[114] Philosopher Martha Nussbaum argues that emotions, which many nonhuman animals have, are inherently intentional, and about something, and embody a way that an animal has of seeing the world; and accepting that animals have emotions is critical both to understanding how they behave and how emotions evolved.[115] Neuroscientist Antonio Damasio claims that brains, human brains, chimpanzee brains, dog brains, are intentional systems, *designed* to be about other things.[116]

More than twenty years ago, Dennett proposed a way to grade intentions. At birth, I may have been a zero-order intentional system, lacking desire and belief. As my mother nurtured me into a first-order intentional system, I achieved both ("I *want* Cheerios"). As I matured into a second-order intentional system, I evolved beliefs and desires about my beliefs and desires and about the beliefs and desires of other intentional systems, such as Mom ("I *want* Mom to *believe* that I am hungry"). I flowered into a third-then higher-order intentional system able, more or less, to grapple with the perplexing questions that arise when my true love says, "Let's not exchange Christmas presents this year." Is she saying that she doesn't want to exchange Christmas presents this year, or does she *want* me to *believe* that she *wants* me to *believe* that she *wants* me to buy her a really nice Christmas present?[117]

What are desires and beliefs? Because beliefs are about what we can imagine, they must be intentional. Some scientists even interpret expectancy-violation tests as showing infants having primitive beliefs, perhaps ones they were born with about how they, others, and the world work.[118] Many scientists speak of these beliefs as involving what babies "expect" or "don't expect," and write that babies show "surprise" or "predict" outcomes, or infants "know" or "understand" things.[119]

Must desires also be "about" something that can be conjured? When I *want* to take Marbury to the Charles River to play, must I mentally represent me, Marbury, and the river bank?[120] Tom Regan, a philosopher, and Henry Wellman, a child development psychologist, think very young children and nonhuman animals can have "simple desires" not "about" anything at all.[121] Some philosophers, such as R. G. Frey, scoff and argue that desires and beliefs require an ability to mentally represent.[122] Frey, writing before the days of expectancy-violation tests and grumpy about what he thought was the unwarranted extension of rights in general, claimed that no one, human or nonhuman, could desire or believe without language.[123] Piaget disagreed, as have most others since. Child developmental psychologist Janet Astington, apparently unable to imagine grumpy philosophers, flatly insists "[n]o one would deny that infants have beliefs and desires. We can see their surprise when something unexpected happens—when a belief is not confirmed—and we witness their rage when a desire is frustrated."[124]

When did Christopher begin to desire and act intentionally? Because babies can be "intentional agents" while unaware of their intentions, the problem becomes complicated. The issue is unimaginably hard even to investigate.[125] Piaget, who sighed that "nothing is more difficult to define than intention," thought Christopher sufficiently demonstrated intentional agency when he could perform an intervening act, such as shoving aside an object to reach his goal, and the more intervening acts he could perform, the more certain we should be.[126] This began at Substage 3 and clearly existed no later than Substage 4. I say "no later" because Piaget's criteria were a lot tougher than those used today.[127] Since nobody accuses Piaget of *overestimating* infant cognitive abilities, it's probably safe to say that by Substage 4, Christopher could desire and act intentionally. Now when did he achieve a self?

Christopher Gets a Self

A self allows an animal to know that the events of her life are happening to her. There are multiple ideas about self and we'll look at several of the most influential.[128] Pioneering psychologist William James thought two "selves" existed. "The self as knower" (the "me-self," the "explicit self," the "objective self") is what we usually think about when we think about ourselves ("Where should *I* go tonight?" "*My* hair's a mess!" "*I* don't feel well"). It requires mental representation and is plainly revealed in mirror self-recognition (MSR).[129] The standard mirror self-recognition test (MSR) developed by psychologist Gordon Gallup, Jr., which I described in Chapter 3, and its variants adapted for other nonhuman species and human infants are widely used as markers for visual self-recognition.[130]

Thirty years on, what it proves to pass an MSR test remains controversial. Some argue it has nothing to do with self-awareness.[131] Some say it shows an understanding of Substage 6 object permanence.[132] Some claim it reveals the ability to imitate; others disagree.[133] Some think it implies a theory of mind; others disagree.[134] And some contend it evidences an ability to understand causality and spatial relations or mentally represent.[135]

What if an animal fails? Human children and apes tend to follow a similar pattern of MSR response. There is no MSR early on. Then it comes and goes. Finally it stabilizes.[136] But how animals learn their bodies correspond to a mirror image is complicated and ill understood.[137] MSR relies upon only one kind of recognition (visual) in a medium (the mirror) that Louis Moses dryly notes "is surely low in terms of ecological significance for most species."[138] Some see MSR as just a milestone in a process that unfolds from the beginning to the end of life.[139] Others think it involves more complex intellectual abilities than self-awareness, for some kinds of self-awareness appear long before MSR. If that's true, MSR flunkers might demonstrate self-awareness in other ways.[140] Some emphasize MSR requires not only an ability to understand the "me-self," but how the "me-self" relates to a mirror.[141] A creature might understand something about her "me-self," but not realize that it's staring back at her from a mirror. Or individuals of some species may not care enough about a body mark to touch it.[142] Or they might not be interested in their faces, which is how some scientists explain MSR failures of humans before the age of about eighteen months.[143]

Sometimes it takes some creative thinking. Monkeys don't usually pass mirror self-recognition tests. But monkeys view stares as threats; they simply don't stare at other monkeys. But if a monkey won't stare into a mirror, how will he learn that the monkey in the mirror is himself? Perhaps, cognitive science professor Marc Hauser thought, if he could be induced to look just a little longer.

Cotton-top tamarins are small monkeys with black faces and a huge thatch of white hair. Hauser's graduate students located bottles of Manic Panic hair dye in trendy Harvard Square, then transformed Hauser's lab into a tamarin salon. Given torn jeans, dirty T-shirts, backpacks, and nose-rings, the tamarins might have blended into any number of gatherings in the Square. Hauser plunked them before a mirror and they sat, seemingly spellbound by their punk green, pink, or blue tops. Some began to compare the mirror movement of their hands and the images of their hands; others used the mirror to check out their butts. Hauser concluded the tamarins demonstrated mirror self-recognition.[144]

James's second, less-complicated self is "the self as known" ("I-self," the "implicit self," or the "subjective self").[145] This is "the self who thinks, feels, and experiences."[146] It understands, however vaguely, that it is living its life. This self is sufficient for practical autonomy; it doesn't require mental representation. Irene Pepperberg thinks a creature who demonstrates object permanence "comprehends the permanence, separateness, and continuity of external objects [and] can also comprehend its own identity as a separate identity among others in its environment, and can thus situate itself in time and space."[147] When does this self appear in humans?

Child psychologist Robbie Case used Piaget's ideas as a springboard to investigate the human infant's first sense of self.[148] Case thought in their first four months of life, infants develop "their first dim glimmerings of [James's] emergent 'I-self.'"[149] Evidence suggests they begin to discriminate themselves from the rest of the world at no later than three to five months.[150] Case thought that by four to eight months, a baby obviously experiences the "I-self" along with its "sense of personal agency," and that the "me-self" is starting to glimmer.[151]

Emotional Selves

Self demands consciousness, and emotions may produce them both.[152] A new scientific field, "affective neuroscience," the neurology of emotion, has appeared.[153] Neuroscientist Susan Greenfield argues that consciousness necessarily entails feelings, emotions are its building blocks, and raw consciousness is the very feel of emotions.[154] After thirty years of studying the relationship between emotion and brains, the pioneering affective neuroscientist Jaak Panksepp thinks self-representation emerged early in brain evolution and resulted in brain processes producing a "primordial self," the "ineffable feeling of experiencing oneself as an active agent in the perceived events of the world."[155] This self is "not a very skilled and intelligent self," but "allows animals to develop into the intentional, volitional, and cognitively selective creatures that they are."[156] Panksepp says, "Descarte's faith in his assertion 'I think, therefore I am' may be superseded by a more primitive affirmation that is part of the genetic makeup of all mammals: 'I feel, therefore I am.'"[157]

Philosopher Martha Nussbaum writes flatly, "Animals have emotions. Few who have lived closely with dogs and apes would deny this; and most researchers agree in ascribing at least some emotions to many other animals as well."[158] How can we know? Nussbaum points to the morally ambiguous, even reprehensible "learned helplessness" experiments that Martin Seligman performed, often by shocking dogs unable to escape, that produced obvious palpable fear, grief, and depression.[159] And she relies on the account of such nonhuman animal emotions as anger, fear, and love that philosopher George Pitcher carefully relates in his *The Dogs Who Came to Stay.*[160] Sure, she says, "(t)here's always room for skepticism. . . . But . . . it's useful to remind ourselves that our attribution of emotion to other human beings itself involves projection that goes beyond the evidence."[161]

Panksepp says "if we consider actions to be valid indicators of internal states in humans, we should also be ready to grant internally experienced feelings to other animals."[162] Marian Stamp Dawkins reminds us that nonhuman animals may express emotions in ways that humans find difficult to detect because we don't recognize what they are doing. "A bear covered with fur is not going to give very much away if and when it blushes," Dawkins says, "and birds, by the constraints of their anatomy, cannot raise a clenched fist or even smile."[163]

Damasio has been labeled "the major living figure in his field."[164] His colleague Raymond Dolan called Damasio's pathbreaking book, *The Feeling of What Happens,* "the first truly compelling neurobiological account of the self."[165] Nearly as impressive has been Damasio's success in convincing colleagues he might be onto something. Psychologist Bruce Charlton wrote in the *Journal of the Royal Society of Medicine* that Damasio had "solved the problem of the nature of consciousness."[166] In a cooler moment, philosopher Thomas Metzinger still baptized the book "a landmark in the interdisciplinary project of consciousness research," and neuroscientist William Calvin tagged its revelation of the transcendent role of emotions and feelings in the experience of consciousness a "really impressive feat."[167] Calvin confided that for years, he dismissed the idea of approaching consciousness from how it *feels.* Only when Damasio demonstrated that consciousness and emotion are inseparably intertwined by highlighting weird neurological syndromes that precisely altered patients' consciousness did Calvin sign on.[168]

Damasio says two basic kinds of consciousness, core and extended, exist. Each has a self. Readers of *Rattling the Cage* will recognize Damasio's embrace of its Number One Theory of Consciousness: identity.[169] This means the biological processes of our brains don't *cor-*

respond to mind processes, "they *are* mind processes."[170] A third self exists. We can't directly know about it because it is a nonconscious protoself.[171] Animal brains incessantly mint mental images of their kaleidoscopic internal states, but outside their awareness.[172] This "collection of neural patterns which map, moment by moment, the state of the physical structure of the organism in its many dimensions" grounds the two conscious selves.[173] These endless wordless stories produce our feelings.

Feelings are not the same as emotions, says Damasio. Emotions are physiological changes and behaviors that physically affect our bodies. Hormones surge. Skin flushes. Body postures change. Hearts race. Anyone who has fallen in love or been audited by the IRS understands. Feelings are the private experience of emotions. Handy things, feelings, for they allow bodies to react automatically at critical moments. That is why so many species evolved them.[174] But feelings and emotions need not occur together. Creatures with just protoselves may feel without knowing they feel or feel without knowing that the feelings they feel are theirs, even as they exhibit emotions that lie beyond their awareness or control.[175] Lacking consciousness, they lack the experiencing of selves.

Organisms constantly generate feelings and produce a flow of mental representations of objects—defined broadly to include a human being, the city of Boston, a toothache, and memories of any of them—that they perceive in the world.[176] Sometimes the protoself's feelings collide with a mental representation of an object. This produces the "feeling of what happens" (hence Damasio's title). Previously unconscious mental images leap into consciousness. This feeling *is* the core consciousness that produces the "I-self" that exists at every instant.[177] It is an activity of indescribable complexity, for a separate core consciousness is created for each of the millions upon millions upon millions of mental objects with which a brain interacts. Each separate act of creation is "the sense of self in the act of knowing," and each results from a symphony of basic brain structures that work without a conductor and are independent of reason, memory, or language.[178]

Any creature possessing core consciousness has feelings and expresses emotion, is awake and attentive, and is able to concentrate, act purposefully, and focus his attention on the world in a sustained fashion.[179] "Individual perspective, individual ownership of thought, and individual agency are the critical riches that core consciousness contributes. . . . The essence of core consciousness is the very thought of you—the very feeling of you—as an individual being involved in the process of knowing your own existence and of the existence of others."[180]

Piaget compared Christopher's early mental images to a slow-motion movie. Christopher might understand each frame, but not the whole movie. Damasio says the core self is created in pulsing scenes that follow each other so quickly they appear continuous.[181] But they aren't. A core self doesn't know how to connect the scenes the brain creates. But within each scene you still know "that it is *you* seeing because the story depicts a character—you—doing the seeing."[182] A human afflicted by a transient global amnesia is stripped of all consciousness but core. She loses all memory of the past and can't anticipate the future. But although she may ask where she is, how she got there, and what she is doing there, she almost never asks *who she is.*[183]

The second, more complicated consciousness, extended consciousness, appears when core consciousness emerges in an animal who possesses large memory capacities, an ability to hold in mind many things about the present and past at the same time, and adequate reasoning ability. Part of one's extended consciousness is a permanent, complex, autobiographical self that blossoms as autobiographical memory as experiences accumulate.[184] It allows one to place the present in the context of an experienced past and possible futures.[185] Language, though unnecessary, makes the autobiographical self soar.[186]

Christopher either possessed core consciousness at birth or acquired it shortly after birth.[187] Damasio speculates that core consciousness was all Christopher had in the early stages of his life. But as he experiences, his autobiographical memory grows and his autobiographical self emerges. Perhaps his developmental milestones result from "the uneven expansion of autobiographical memory and the uneven deployment of the autobiographical self."[188] Damasio suspects that when Christopher was twelve months old, objects entered his mind, were processed by his core self, and exited, each "known by a simple self and clear on its own, but there is no large-scale relation among objects in space or time and no sensible connection between the object and either past or anticipated experiences."[189]

Neisser's Five Selves

Psychologist Ulrich Neisser developed an influential theory of human self, grounded in the work of the ecological psychologist J. J. Gibson. Neisser argues that human self is not a single entity. Instead, humans have access to five kinds of information about themselves. Each concerns a different aspect of self and each creates a different self.[190]

The most basic is the "ecological self." This develops as one lives, experiences, and perceives one's physical environment, though evidence from development of blind children suggests that cognition, not just perception, may play some role.[191] Perceiving the world is perceiving oneself.[192] In conflict with Piaget's idea that children are born into a world from which they cannot differentiate themselves, Neisser thought the ecological self probably exists at or near birth and becomes established within the first few months of life.[193]

The ecological self is purposeful; it involves "being aware of itself as a choice-making individual."[194] It monitors itself and directs its movements toward specific objects to achieve its goals. It can repeat bodily movements, examine and explore its body, and differentiate itself.[195] It can distinguish between itself and its goals, situate itself in relation to its goals, and perceive how effectively it can attain its goals.[196] It isn't self-consciousness, but simple awareness.[197]

An ecological self is not an all-or-nothing entity. A human fetus has no ecological self, but an eight-month-old human has a developed one. A continuum exists between fetus and infant.[198] Any creature who systematically explores or uses detours, or reaches for objects, has some ecological self.[199] Neisser doubts that flies, being stimulus-response machines, have one.[200] Philippe Rochat argues that the ecological self is not confined to humans but is a property of any biological system that perceives and acts to attain goals, and probably includes every mammal and bird.[201]

Neisser's second self is the "interpersonal self," revealed through apparently unreflective social interactions. A being is aware of herself as a social agent. Human babies may have an interpersonal self almost from birth, if their limited ability to imitate, a goal-directed intentional activity, is an indicator.[202]

The "conceptual self" appears around the first birthday, when one can mentally represent one's characteristics.[203] When a human knows and remembers her life, realizes it's an unbroken story, communicates about it, and projects it into the future, she possesses the fourth "temporally extended self," which appears during the fourth year.[204] The fifth, "private self," emerges when a child understands that her conscious experiences, thoughts, and dreams are her own; it occurs at the fifth year, or beyond.[205]

The experience of pain shows how different selves experience. A human infant, or nonhuman animal, can probably experience pain without experiencing self: "Even when it produces reflex responses or disrupts ongoing behavior, pain may dominate conscious experience in an unstructured global way without being attributed to anything at all like a self, or indeed to anything at all."[206] When I reflect on what pain might mean—do I have cancer?—I use my conceptual self. When I think about the pain, how much it hurts, I use my private self. I may experience my throbbing leg as part of my ecological self. But pain may disrupt my functioning, especially in its earlier, more fragile stage. That is why Neisser thinks a baby is unlikely to stick a hurt finger into her mouth, but an older child will.[207]

Practical Autonomy at Last!

Present evidence suggests that at birth, Christopher lacked practical autonomy and his birth self should be placed in Category Four. If he wasn't born with core consciousness, with Damasio's "I-self," it soon appeared. Between four and eight months, his Jamesian "I-self" emerged, and his ecological and interpersonal selves were settled by eight months. Then, or perhaps at Substage 4, he began to desire, perhaps to believe. His intentionality, which first appeared at Stage 3, was established. No later than Stage 4, Christopher probably desired and became an intentional agent. Now he was in Category Two, moving from 0.51 to 0.89. Sometime after his first birthday, as his language and conceptual self appeared and his autobiographical self began to flourish, he ascended to the 0.90 required for Category One, then to 1.0, or full practical autonomy.

FIVE

Honeybees

Evolutionary Bookends: From Honeybees to Chimpanzees

In dozens of interviews and public talks after *Rattling the Cage* appeared, I was asked, usually facetiously, where I would "draw the line." Monkeys? Cows? Dogs? Snakes? Frogs? and (wink, wink) what about insects? Had I answered this last in the affirmative, scorn would have been heaped upon me. Worse, my answer would have been used to counter the exceedingly powerful arguments for the dignity-rights of chimpanzees and bonobos. Even Donald Griffin appears daunted by the prospect of insect rights: "[W]e can scarcely manage to avoid the injury and suffering of all insects. . . . Whatever the approach, there surely comes a point . . . at which for practical and ethical purposes it is reasonable to draw a line."[1]

I chose the honeybee as an example of a Category Four animal having an autonomy value below 0.5. To my amazement and horror, the more I learned about the mental abilities of honeybees, the more certain I became that bees were Category Two animals. The questions was: how high?

Honeybees, *Apis mellifera,* and I are very distant evolutionary relatives. For answers on how distant, I turned to "Mr. Darwin," a pseudonym for an evolutionary biologist who prefers anonymity. He said we're both in the animal kingdom and that's about it. Our common ancestor lived between 550 million and 750 million years ago; he can't be more precise.[2] We'll call it 650 million years. In contrast, I last shared a common ancestor with the chimpanzee whom I met in Uganda, Big Brown, between 5.5 million and 6.5 million years ago.[3] Let's call it 6 million.

How long ago is 6 million years? What about 650 million? How do they compare? Eons are hard to grasp. I chatted about the problem first with Cam Muir, a postdoctoral research associate at the University of Hawaii. He suggested we convert the time into geographical distances, perhaps three feet per generation.[4] This strategy works well with apes and other primates because we can make reasonable guesses about their generational lengths. But as we wade into the past, we know less and less about the length of generations. Eventually we know almost nothing at all.

Try this. Hold your breath. We'll let one second of breath-holding stand for 3 million years. Take a breath and hold it for two seconds. One. Two. That is how long ago you and Big Brown had a common ancestor. Let's turn to honeybees. Take a breath. Now hold it for three minutes and thirty-seven seconds. That's how long ago you shared a common ancestor with a honeybee.

My brain weighs about three pounds. Big Brown's weighs about a pound; nobody's really sure how many neurons are packed into either. Guesses often range between ten to the tenth or ten to the eleventh power.[5] I have about three times the number of brain neurons Big Brown has and a similar brain structure, which is not surprising because we were the same animal just two evolutionary seconds ago.[6] But the volume of a honeybee's brain is only about one cubic millimeter, and its structure has almost nothing in common with any vertebrate brain—just what we might expect of brains that diverged so long ago.[7]

A honeybee's brain holds less than 1 million neurons; some say about 960,000, others about 850,000, and half belong to the optic lobes. A bee's brain is therefore about a millionth the size of Big Brown's brain. If an entire honeybee brain was substituted for each honeybee neuron, the resulting brain would be about the size of Big Brown's.[8]

Bee neurons appear to work in two major ways. Some are dedicated to one specific function, just as a dedicated fax machine does one thing, send and receive faxes. The mushroom bodies, much more complex systems of about 170,000 crowded neurons, are examples of multiple parallel neuron architecture; this arrangement apparently accounts for the remarkable cognitive abilities that, as we will see, bees developed that allow them to act so flexibly.[9]

Honeybee Learning and Memory

Honeybees possess a mental ability many think is strictly the domain of primates: the ability to recognize "sameness" and "difference."[10] Bees were taught to solve "delayed match-to-sample" and "delayed non-matching to sample" problems. In the first, they matched two identical abstract symbols. The second required them to understand when two symbols differed. Experimenters thought honeybees learned "sameness" and "difference" only when they could solve both problems by using symbols not explicitly taught. Here's how it worked.

Honeybees were trained in a Y-shaped maze. A bee could buzz into a hole cut from the laboratory wall into the maze. She might see a vertical or horizontal sign at the entrance, or a blue or yellow sign (the honeybee has color vision). She could fly into a "decision chamber" to encounter the arms of the "Y." A sign identical to the one at the entrance was posted on one arm, a different sign on the other. If the bee chose the arm with the matching sign, she was given a sip of sucrose. Bees learned this. They also learned to match odors to choose the sign that didn't match.

Honeybees displayed such "excellent learning capacity" that researchers began giving them tests normally reserved for vertebrates.[11] In 1997, Randolf Mendel, professor at the Institute for Neurobiology at the Free Institute in Germany, wrote that "[h]oneybees collect nectar and pollen from flowers, not randomly, but with a systematic search strategy. The most important component of this strategy is the learning behavior of the individual bee, and the long lasting memory which results from a fast and effective learning process."[12] By 1996, another leading researcher could write that "[t]he performance of honeybees in a variety of experiments patterned after experiments on learning in vertebrates is strongly reminiscent of the performance of vertebrates."[13]

Louis Herman, who has taught simple communication systems to Atlantic bottle-nosed dolphins for thirty years, argues that memory is the basic cognitive building block.[14] Honeybees have lots of it, defined by researchers as "an animal's capacity to retain acquired information and to use it for future behaviour."[15] For a glimpse of what Herman means, imagine that you have lost your memory.

Menzel and Martin Giurfa have found five memory phases that reflect the sequence of events during a honeybee's typical nectar foraging trip. Early and late short-term memory are triggered by a single learning trial. Early and late long-term memory require more than one trial, but late long-term memory lasts a honeybee's lifetime (weeks or a few months at most) after just three learning trials.[16] Honeybees recall odors.[17] And whatever rules they use in foraging, they must remember.[18]

Some days while driving my car, I turn on WODS, an "oldies" radio station that plays songs from my college days in the late 1960s and early 1970s. I like the songs, but something complex can happen as I listen. I may associate a song with the smell of my girlfriend's perfume at the time I heard it, all those years ago. It's called "cross-modal associative recall." To some extent honeybees have this type of recall, and here's how scientists know that.

Researchers trained bees to fly into a chamber scented with mango or lemon. The bees could choose an exit marked by either yellow or blue. They received a reward if they exited yellow after smelling mango or exited blue after smelling lemon. Because they quickly associated each scent with the color, when the bees flew into a mango-scented chamber the following day, they usually chose the yellow exit. If they flew into a lemon-scented chamber, they exited blue. Sometimes the bees hesitated and turned to hover before the scent-dispensing vial again before choosing. They could also be taught to reverse the associations and associate mango with blue and lemon with yellow. They were grouping in their memory stimuli the rewards obtained through different senses, and recalling one when presented with another.[19] Not exactly recalling the scent of Mary Jo's perfume at the sound of "Sergeant Pepper's Lonely Hearts Club Band." But close.

Honeybees use memory to navigate. It's not clear whether they have a "cognitive map," complete with symbols, in their heads, or whether they simply remember the landmarks of routes flown. Marc Hauser thinks the former is more likely. Otherwise they would have to store an enormous number of images in their little bee brains to navigate as well as they do. But Hauser is also impressed by how honeybees react to the following situation.

Princeton biologist James Gould chose a hive that sat near a lake. Each day he trained forager bees to come to a boat filled with nectar. He didn't allow the bees to return to the hive until he placed the boat squarely in the middle of the lake. Foragers could then come to the boat, gorge with nectar, and return to the hive. There they waggle-danced. (Later I'll explain how a waggle dance communicates nectar's distance and direction.) Any hive bee paying attention would receive a message that nectar is located in the middle of the lake! Almost no bees fell for this and departed for the boat, apparently not "believing" the waggle dancers' message. Attempting to rule out the possibility that the results demonstrate only that bees don't like to fly over water, Gould placed the boat on the opposite shore and trained the bees to come there. When they returned to dance, many bees immediately flew straight across the lake to the boat.[20] Hauser concluded that honeybees not only possess symbolic cognitive mental maps but have a mental tool for skepticism![21]

Dancing Bees

In 1973, Karl von Frisch won a Nobel prize primarily for his discovery that honeybees possess far and away the most complex and sophisticated system of communication in the animal world, after human language. He called theirs a "dance language." Honeybees' dances are most frequently, but not always, about food, how far away it is, in what direction, and how good it is.

Honeybee communication involves signals and cues. A signal is a deliberate act of communication shaped by natural selection; a cue is unshaped behavior that the bee doesn't use deliberately to communicate.[22] At least seventeen signals and more than thirty cues have been detected.[23] Five well-studied signals deal with food-finding. We'll examine them because there is little doubt that honeybees, with less than 1 million brain neurons, use symbols to convey complex information understood by other bees.[24] And keep in mind that a foraging honeybee's life is short. She may seek food for only ten days before she dies; her entire life may not be much longer.[25]

Food dances don't occur when a hive is humming and individual bees can obtain enough through individual searches. It's when scarcity strikes that dancing starts. Older foraging bees search out food, then return to dance to tell others what they found and where they found it.[26] They "round dance" when they discover food relatively near the hive, "waggle-dance" if it's relatively far away.

In a round dance, bees circle one way, then the other, and keep alternating. Where they reverse direction they appear to convey information about the direction in which food lies, ever so subtly, by jagging to the side.[27] As the distance to food lengthens, a round dance becomes a waggle dance. The more distant the food, the longer "wagging runs" bees perform. A wagging dance cycle involves wagging the abdomen thirteen or fourteen times a second, while quickly walking a line that can reach a centimeter if food is half a mile away.[28] The fewer the turns, the longer the dance, the farther the food.[29] At the cycle's end, bees return to where they began to waggle and do it again, and then again.

By tricking them, scientists showed honeybees measure distance according to what they see. They trained bees to fly through tunnels painted with horizontal stripes intended to cause the bees to underestimate distance; to make them overestimate the distance, they were sent through tunnels painted a checkered pattern intended to give them a sense of moving past many objects. If the hypothesis was correct, bees flying past stripes would waggle-dance a shorter time than the distance warranted and longer than the distance through the checkered tunnel warranted. And that's what they did.[30] The researchers then tricked the bees again. They trained them to fly through checkered six- and eight-meter tunnels ending three meters from the hive. After the bees waggle-danced, the scientists fanned out to see where the foragers went. Sure enough, they flew to a food source seventy-two meters away, not eleven. The bees had communicated the distance they felt they had gone, not the actual distance.[31]

The direction of the wagging run tells where food is in relation to the hive.[32] The sun is the usual touchstone, though bees may also use polarized light in a way that correlates with the sun's position. When food lies directly toward the sun, bees waggle-dance straight up the side of the hive. When it lies directly away from the sun, they waggle-dance straight down the side. If the food lies at an angle of forty-five degrees, the dancer moves in a line forty-five degrees right of an imaginary line drawn straight up and down.[33] It's not precise. Foragers may arrive within fifteen to twenty degrees of the indicated direction and 10 to 15 percent of the distance.[34] Try to communicate distance and direction with a waggle dance and you won't do any better. Waggle dancers also convey how desirable a food source is. The better the food, the faster they return to their original positions.[35]

The third of the four signals honeybees use is the "shaking signal." One bee actually shakes another and the victim crawls toward either the hive entrance or the dance floor to be recruited by waggle dancers to forage.[36] In this way, the number of foragers is increased.[37] The fourth is the "tremble dance." A forager who has located a rich food source may return and find the bees to whom she would normally transfer her nectar are busy as, well, bees.[38] This dance is so frantic that von Frisch thought it was "as if they had suddenly acquired the disease of St. Vitus's dance," a spastic nervous disorder of the limbs.[39] One noted researcher, Thomas Seeley, concluded that tremble dances are meant to communicate: "I have visited a rich nectar source worthy of greater exploitation, but already we have more nectar coming into the hive than we can handle."[40] They trigger at least two responses. Bees within the hive interpret tremble dances to mean "I should switch to the task of processing nectar," for that's what they do. Bees outside the hive interpret it to mean "I should refrain from recruiting additional foragers to my nectar source," for they may stop waggle dancing until tremble dancing ceases.[41]

Last, honeybees use a "stop signal."[42] It resembles the shaking signal as it is directed toward a particular bee. The shaking signal attempts to make a bee prepare to forage, but the stop signal has the opposite effect. Usually a tremble dancer rams a waggle dancer with her head while emitting a piping sound. Or sometimes a waggle dancer, or even a member of the bee audience, will do the ramming. No matter who does it, the waggle dancer usually stops dancing. So it appears that the purpose is to stop the recruitment of additional foragers.[43]

Honeybees don't just dance about food.[44] Dances tell where water to cool the colony or certain waxy materials are located. Hives can also become crowded.[45] More queens develop, and the hive prepares to swarm. Older foragers stop seeking food and, for the only time in their lives, search for new cavities in which to create a hive. At the old hive, bees cling in a ball to plants or to the hive itself. Searchers waggle-dance on the very surface of the swarm, but now their waggles convey the distance and direction, not of food, but of cavities they have located. And they convey something of the quality of their finding; small, wet, insect-filled cavities bring forth dances few and weak; but larger cavities that are dry and dark ignite comparative whoopie.[46]

By marking individual bees in the 1950s, University of Würzburg zoology professor Lindauer noticed that a large number of scouts danced information about many competing cavities and that something remarkable often occurred. Over several days, inferior sites lost supporters. It's not that scouts just stopped advocating for a site. At least some who danced in favor of an inferior cavity actually visited an allegedly better cavity, then switched allegiances!

In 1999, Lindauer's work was repeated, using improved technology. Researchers marked 9,258 honeybees, in three swarms, and videotaped all dances performed over the several days it took for each swarm to decide where to move. Many sites winnowed to one, and within about an hour of the time nearly every dancer was dancing the same dance, the swarm flew to that site. The researchers found that 305 dancers who initially favored a site that wasn't chosen changed their dances to the site that was. A few changed to other sites. Most abandoned the dance floor. Forty-two percent of dancers who danced in support of the site finally chosen kept advocating for it and the rest sat out.[47]

Griffin finds it "reasonable to infer that under these conditions bees are thinking about cavities and are able to change their 'allegiance' from one they have discovered themselves to a better one they have learned about as recipients of symbolic information from the dances of one of their sisters. . . . [I]ndividual honeybees can, so to speak, change their minds as a result of exchanging information through the communicative dances."[48]

Are honeybees conscious? Griffin says, "If we accept communication as evidence of conscious thinking, we must certainly grant consciousness to honeybees."[49] Do their tiny brains produce sentience? Apparently so.[50]

How Do We Know That Nonhuman Animals Want?

That human babies without language "want" things will surprise no parent. But how do we know that languageless nonhumans "want" things? We should be confident that if human infants pass Piaget's rigorous requirements for any developmental stage, they have the cognition he claimed. That means humans who reach Substage 4, perhaps even Substage 3, have practical autonomy. The closer genetically and evolutionarily an animal is to us, the more confident we can be that if she demonstrates Substage 4—perhaps Substage 3—skills, her underlying mental processes will be similar to ours and she has practical autonomy, too.

Do Honeybees Have Practical Autonomy?

Both William James's and Antonio Damasio's "I-self" are as ancient as emotions and feelings. Just how old, and in what animals can they be found? Calvin and Damasio agree that bonobos and chimpanzees have autobiographical selves.[51] Damasio says the brain structures responsible for core consciousness exist in "numerous nonhuman species" and argues that the lower rungs of extended consciousness are not reserved for humans and apes.[52] "The elaborate and time-consuming behavior of a troop of baboons concerned with choosing the place where they should drink on a given day" is evidence that baboons share it.[53]

Biologist Marc Bekoff writes "current research provides compelling evidence that at least some animals likely feel a full range of emotions, including fear, joy, happiness, shame, embarrassment, resentment, jealousy, rage, anger, love, pleasure, compassion, respect, relief, disgust, sadness, despair, and grief."[54] But which ones? All mammals, Bekoff says, "probably experience similar types of basic emotional and motivational feeling states."[55] Maybe not just mammals. Susan Greenfield thinks all animals have emotions, for they are "a basic property of the brain."[56] Marc Hauser concurs: "Few students of animal behavior would argue that a dog, or any other animal, lacks emotion."[57] For Jaak Panksepp, it "does *seem* self-evident to most observers that animals experience emotional states"; evidence that the brains of all mammals, perhaps all vertebrates, and possibly some invertebrates, have basic emotional operating systems and an "affective consciousness" of similar origin that operates in similar ways is "overwhelming."[58] Neurologist Gerald Edelman believes that in most mammals, some birds, and maybe some reptiles, a primary consciousness can be found that allows them to think, have present mental images, and an "I-self." Because they lack a higher-order secondary consciousness, they are not conscious that they are conscious and have no "me-self."[59]

Of all the animals I have studied, the honeybee is the most difficult to place. They are an extraordinary puzzle. I initially chose honeybees as an example of a Category Four animal. I soon learned how wrong I was. Just how wrong was that?

Donald Griffin quotes Carl Jung's response to von Frisch's work: the message of the dancing bees, Jung said,

> is no different in principle from information conveyed by a human being. In the latter case, we would certainly regard such behavior as a conscious and intentional act and can hardly imagine how anyone could prove in a court of law that it has taken place unconsciously. . . . We are . . . faced with the fact that the ganglionic system apparently achieves exactly the same result as our cerebral cortex.[60]

If Jung is correct, we must consider placing honeybees in Category One and granting them basic legal rights. At this time, however, I don't think the evidence supports Jung.

Jung represents one extreme of opinion about honeybee dances. At the other extreme, Adrian Wenner and Patrick Wells have long argued that the dances are not symbolic communication at all, but one gigantic delusion, perhaps even a scientific fraud, on the level of N-rays, polywater, water with memory, and cold fusion.[61] This is a serious charge. N-rays, for example, are not real, and their alleged existence was famously debunked when a physicist visited the laboratory of one prominent advocate. In the darkened room, he surreptitiously pocketed the alleged source of the rays right in the middle of an experiment, which affected the results not one whit![62]

In their book, Wenner and Wells frequently invoke William Broad and Nicholas Wade's book, *Betrayers of the Truth: Fraud and Deceit in the Halls of Science,* in which even ape language is declared self-deception.[63] Their thesis is that honeybees follow odors, bee dances are poppycock, and the decision to embrace either the dance or the odor argument is not really scientific.[64] In their view, honeybee researchers "easily become emotionally attached to the subjects, consider them superior creatures 'second only to humans,' and imbue them with humanlike capabilities. . . . An almost religious fervor sets in."[65] I think Wenner and Wells are no more correct than was Jung.[66] The truth lies between.

Piaget's child development scheme is of no help with honeybees. It will become more helpful when we encounter animals evolutionarily far closer to us. Michael Tomasello and Josep Call, for example, find Piaget's stages of sensorimotor development "extremely useful" for studying some aspects of primate cognition.[67] As with many animals, bees don't have the arms and legs needed to look good on Piagetian tests. Some species investigate objects not by manipulating them but by looking at them, smelling them, or echolocating them.[68] That's why Tomasello and Call don't like Piaget's emphasis on cognitive development through manipulation of objects, for "many nonhuman species may have cognitive skills that simply do not manifest themselves in the manipulation of objects. Since manipulation of objects is the behavioral expression that Piaget used almost exclusively as an index of cognitive functioning in human infants . . . the cognitive skills of some species will be seriously underestimated."[69] Think of snakes, parrots, bats, dolphins, and fish, and it's not hard to imagine honeybee difficulties.

The subtleties of honeybee cognitive abilities can be detected only by scientists who dream up creative ways of testing minds and bodies that may be strikingly different from theirs. Some have already devised ways of asking certain animals questions they can answer in ways we understand. For many years in England, egg-laying hens have been crammed into battery cages. In the 1960s, a government committee recommended that battery cage floors be constructed from heavy, not fine, wire. Committee members reasoned that thicker wire would provide better support for the hens' feet. Then two doubters asked the hens. They constructed cage floors from thick wire and cage floors from thin, and allowed hens to stand wherever they wished. The hens preferred the thin, apparently because the strands acted like a snowshoe to support their weight.[70]

Languageless animals may even tell us what they want when we don't ask. One scientist investigating the effects of inhaling cigarette smoke pumped it through a glass tube into barren glass containers that held unfortunate mice and hamsters captive. Many responded by damming the tube with their feces to block the smoke, sometimes to the point of asphyxiating themselves.[71]

In the 1980s, Marian Stamp Dawkins refined a technique for asking languageless animals how *much* they want something. Marc

Hauser admiringly allows that Dawkins's work gives us the opportunity "to look at the psychology of desire by exploring the economics of *choice.*"[72] Here's how it works. A scientist imagines what an animal might want. Then she decides how to place increasingly difficult obstacles in his way. How hard he works to obtain what he wants suggests how much he wants it.

When a decision was made in England to shave the costs of battery egg production by removing the chipped wood that lined cage bottoms, animal protection groups reacted angrily, arguing that hens need chipped wood to scratch. To find out, Dawkins built a cage with two cells and a door between them that a hen could open. The floor of just one cell was lined with chipped wood. When she placed a hen in the other cell, the hen would open the door and enter the cell lined with chipped wood. Hens clearly wanted their cage floors lined with chipped wood. But how much did they want it? Dawkins made it harder and harder for them to open the door. The hens responded by trying harder and harder to open it.[73] This time the animal protectionists were correct.

The excellent learning capacity and memory of honeybees, their likely symbolic cognitive map, symbolic communication code, sophisticated communication system, and apparent ability to engage in rudimentary thinking puts them above Categories Three or Four. They don't qualify for Category One. That leaves Category Two. But how high up in Category Two should they be placed? If they were vertebrates, I would place them at 0.75, even 0.80. This would not be an exact fit, for it is unclear whether they have even ecological selves. I suspect they may. But I don't know. The vastness of the evolutionary distance between honeybees and humans makes it difficult, nearly impossible, to imagine what they experience, even whether they are conscious. In 1956, R. E. Snodgrass, a honeybee expert, stated he didn't know whether honeybees have any degree of consciousness.[74] In 2001, M. E. Bitterman, who has also spent his professional life studying honeybees at the University of Hawaii, responded to my query: "If you're asking whether honeybees are conscious in the sense that (based on personal experience) we assume consciousness in other human beings, I must admit to ignorance."[75] On the other hand, Donald Griffin says that "[i]f we accept communication as evidence of conscious thinking, we must certainly grant consciousness to honeybees."[76]

Such profound disagreement is a major reason why I assign honeybees an autonomy value of 0.59. On present evidence, I do not think that honeybees can be said to possess practical autonomy. If they do, that doesn't mean that other insects do, too. But it's a closer call than I imagined before I began to study the evidence. As scientific investigation continues, the evidence for practical autonomy may be weakened, or strengthened, for as one of the world's most famous and experienced honeybee experts pointed out decades ago, "You never can tell with bees."[77] If it is strengthened, their entitlement to basic rights will have to be revisited.

The physical, economic, and psychological obstacles to dignity-rights for honeybees are overwhelming. For the year 2000, the United States Department of Agriculture reported that 2.63 million honeybee colonies were used for the commercial production of 221 million pounds of honey worth $126 million.[78] At a conservative 35,000 honeybees per colony, that's almost 100 billion honeybees.[79] The harvesting of honey as it is done in commercial production kills bees.[80]

The political and psychological problems inherent in granting any rights to nonmammals, nonvertebrates, and especially to some insects, perhaps even the kind of legal protection given them by anticruelty statutes, may be the biggest hurdle. Several American states have refused protection to nonhuman animals such as fighting gamecocks on the ground they are not "animals" within the meaning of an anticruelty statute, and I am unaware of any court that has protected an insect.[81]

SIX

Alex

When I called Irene Pepperberg to make an appointment to meet Alex, she told me that I could find him at MIT's Media Lab in Cambridge, Massachusetts. I followed her directions to a large dark computer-crammed room. A window ran the eighteen-foot length of the room. Popping my head inside the doorway, I saw three African Grey parrots and five young humans dressed in jeans and T-shirts, islands of "brick-and-mortar brains" in a sea of virtual intelligence. One young man asked my business.

"I'm looking for Alex."

He briskly directed me out.

"I know Irene," I said, calling after him.

He kept walking.

"I have an appointment."

He kept walking.

We finally stopped before a small office. Pepperberg was chatting inside. I waved and she waved back. The young man relaxed.

"You'd be surprised who comes looking for Alex," he said, and led me back to the room.

One parrot, half his red tail feathers missing, was standing one-legged atop a three-foot-high heavy-duty steel cage. He was gripping a paper coffee cup with his talons and pulling at it with his beak. A second, sleek with long red tail feathers, preened on another cage. He ignored me. The third, scruffy, lacking tail feathers, stood on a third cage. He cocked his head and gazed at me with an unblinking left eye.

All three cages were filled with shredded things, shredded coffee cups, a shredded Kleenex box, shredded corks, and a shredded "Annie's Spinach Feta Pocket Sandwich" box, as well as a bowl of fresh corn, banana pieces, grapes, broccoli, green peppers, parrot chow, and water.

"Which one's Alex?" I asked.

The young man pointed to the scruffy tail-featherless one. Alex was staring at me, wagging his butt. Pepperberg had suggested that I might bring colored pasta as a gift. I produced the bag and showed it to Alex.

"That one's Griffin." The man pointed to the cup-shredder. "The other one's Wart. Alex is interested in you. Would you like to hold him? Stay alert."

Alex was having such a bad day he'd bitten Pepperberg on the thumb. With the unconcern of the wholly ignorant, I stretched forth my hand. Alex strutted up my arm then back down. He danced back and forth, with increasing speed, over the next five minutes.

"That's courtship behavior," the man said. "Watch out. He might regurgitate."

"He already did," a woman said. "I don't think he'll spit it on you, though."

He didn't. He just kept dancing. I knew Alex occasionally courted Pepperberg's male students.[1] Being the object of a same-sex interspecies flirtation with a parrot whom I had just met—and before all these people—caused a fluttering of confusion. Happily, Pepperberg appeared, looking professional in a plaid blue skirt and jacket. A pair of sunglasses balanced on her head. Her left thumb was wrapped in a blood-stained Band-Aid.

"He's not cooperating today," she said. "Let's see if he wants to work."

She held out her hand. Alex jumped on, then hopped to a long half-gnawed wooden perch, across from Griffin. Pepperberg rummaged through plastic bags filled with magnetic colored letters and numbers of the kind I stick to the refrigerator for Christopher and Siena.

"Shower," Griffin said.

Pepperberg turned.

"You hate showers."

She looked at me.

"We're learning phonemes. To help them get the idea that words are made of sounds."

She arrayed a green "u," a red "t," yellow "s," purple "n," green "or," red "i," and orange "sh" on a worn, circular piece of cardboard duct-taped along the margins, then showed the board to Alex. From half a foot away, he cocked his head and studied the letters.

"Okay, what sound is yellow?" Pepperberg asked.

"Ssss," Alex said.

"That's right!"

"Wanna nut," Alex said.

"Where are the nuts?" Pepperberg asked.

Alex's head was already buried inside a cup.

"In the cup," someone called. "Near Alex."

"What sound is orange?" Pepperberg asked.

"Shhh," Griffin said.

Pepperberg glared at Griffin.

"No, Grif, not yet! Alex, what sound is orange?"

"Ssss."

"No. Try again."

"Shhh."

"Right."

"Wanna nut."

"No. Only when you get it on the first try. Now, what sound is the green 'u'?"

Alex was silent.

"What sound is green?"

Silence.

"Come on, Alex! He's hardly done anything since he bit me," she said to me. "What sound is green?"

"Ssss."

"Alex! Sometimes he does this. We had visitors a few months ago. He kept asking for a nut. Nobody paid attention. So he says, real loud, 'NUT! Nnnn . . . uhhhh . . . tttt! That got everybody's attention! But will he ever do it on a test?" She rolled her eyes. "What sound is green?"

"Oooo," said Alex.

"Right!"

"Wanna nut."

"Okay."

"Want corn," Griffin said.

Pepperberg ignored him.

"What color is 'or'?" she asked.

"Green."

Alex seized a nut.

"What sound is yellow?"

"Ssss."

"Right. What sound is red?"

"Eeee."

"Right!"

She showed the array to Wart.

"What sound is yellow, Wart?"

"Ssss," said Alex.

"No, no," Pepperberg said. "Let Wart try. He's just learning."

Young Wart, a couple of years old and a relative novice, got nowhere. Pepperberg dumped the letters. She filled the cardboard with numbers, a blue "1," yellow "2," red "3," green "4," black "5," and orange "6." People were filtering in and out of the room. Suddenly there were eight humans. Everyone was tapping on a keyboard except for Pepperberg and me.

"Okay, Alex, what number is orange?"

"Six."

"Right. What number is green?"

"Four."

"Right. What color is '1'?"

"Yellow."

"No. Yellow is '2.' What color is '1'?"

Alex ignored her.

"Pay attention!"

He was unmoved.

"Alex, what color is '1'?"

"Green."

"Sometimes he gives everything but the right answer. Come on, Alex. What color is '1'?"

"Yellow. Wanna nut!"

Pepperberg was exasperated.

"Let's take a break."

She opened one jar of organic pureed peas, another of rice and lentils, and offered small quantities in a spoon. The parrots scooped the food into their mouths with large curved beaks. I began idly to fiddle with the joystick of a nearby computer. Every time I pulled, a half-inch red ball appeared. I could make a line of red balls go up and down or back and forth. When Pepperberg finished feeding the parrots, she carried Alex to a perch before the screen and used the joystick to write an "S" in red balls, six inches high and wide.

"Alex. What sound is this?"

He cocked his head, but said nothing for a long time.

"He's never seen this before," Pepperberg explained to me.

"Ssss."

Finally, two assistants walked over. Pepperberg drew a huge "H" beside the "S."

"What sound is this?"

Alex didn't respond for thirty seconds.

"Shhh."

"Hey," an assistant said, "he can see it."

Pepperberg quickly erased the "SH" and drew a big red "I."

"What sound is this?"

"Eeee."

Another assistant joined our group. People were getting excited.

"Try another!" one said.

Pepperberg drew a huge "N." "What sound is this?"

The answer came quickly.

"Nnnn."

Pepperberg looked at me. "He transfers to the screen! I'd say he's redeemed himself. For today."

A Revolutionary Advance

Is what Alex uses "language"? Experts are divided over whether even infant Christopher used language. If nothing else, language is communication that uses symbols.[2] An animal communicates when she intentionally behaves so that another senses it and reacts.[3] Communication might be with words or gestures. Psychologists consider a gesturing human infant, unable yet to speak, communicates when making eye contact at the same time she gestures to indicate that something has not been accomplished.[4]

Months before, I had been astonished to hear Pepperberg refer to Alex's "language." At least I *think* I heard it. In two decades of scientific reporting, Pepperberg has unfailingly written that Alex speaks in a "human-based code," a "two-way communication code," a "code based on English speech"; that he engages in "human-based communication" or uses "English labels interactively with human trainers."[5] But Pepperberg has never said "language." Over lunch, I triumphantly repeated what I'd heard her say. She stopped twirling spaghetti, smiled, and pleasantly denied it.

A reviewer for *Nature* of Pepperberg's magnum opus, *The Alex Studies,* wrote, "To misquote Ludwig Wittgenstein, what would a parrot tell us if it could talk? Not a lot seems to be the answer."[6] Pepperberg has no time to argue about whether Alex uses language or to compare his linguistic skills to those of humans. She has never cared to investigate the extent to which Alex can use a human language. She concedes "comprehension of a language-like code, although clearly indicating some complex processing ability, does not necessarily imply comprehension of the full complement of elements that constitute language."[7] She is interested in Alex's "communicative competence," his ability to convey intent and respond to the intent of others, and has spent her entire career opening a communication channel wide enough to allow her to study Alex's general cognitive abilities.[8] If the hallmark of intelligence is the ability to transfer skills learned in one area to another, Alex is intelligent.[9]

Donald Griffin understands what she's doing: "Pepperberg makes no claim that Alex has learned anything approaching the versatility or complexity of human language; but he does seem to have demonstrated some of the basic capabilities that underlie it."[10] Alex has been taught about one hundred words, mostly the names of common objects, colors, and numbers. I suspect the *Nature* reviewer, if confined to what Alex has been taught, wouldn't be a sparkling dinner partner either. Even so, Alex shows the turn-taking that marks communicative competence, and a simple syntax.[11] Sometimes he practices his words when alone.[12] Griffin understands the implications of Pepperberg's work: it "constitutes a truly revolutionary advance in our understanding of animal mentality, comparable to von Frisch's discovery that honeybees use symbolic gestures to communicate to their sisters."[13]

In her data interpretation, Pepperberg can be irritatingly conservative; but when she does draw a conclusion, it is nearly unassailable. In 1998, the *Boston Globe* reported her opinion that Alex and other African Grey parrots "have cognitive abilities comparable to a four- or five-year-old child."[14] Two and a half years later, the same newspaper quoted her statement that "parrots reason, comprehend, and calculate at the level of a four-year-old child."[15] She is not the only one who believes this. In *The Parrot Who Owns Me,* ornithologist Joanna Burger, who has studied parrots for years and shares her domestic life with Tiko, a red-lored Amazon parrot, writes that Tiko's "behavior and intelligence were remarkably like a precocious three-year-old's."[16]

These are remarkable assertions. According to "Mr. Darwin," Alex, Tiko, and I last shared a common ancestor between 300 million and

360 million years ago.[17] We'll call it 330 million, which means, at 3 million years a second, you must hold your breath for one minute and fifty seconds to appreciate it. I asked Pepperberg whether, despite the evolutionary distance between Alex and me, she believes what she told the newspaper. She said she does.[18] Then so do I.

Bird Brains

An intelligent mammal has a large cortex. But walnut-brained birds with almost no cortex can also be highly intelligent. Their grey-mattered striatum produces it.[19] Pepperberg contrasts mammal and bird brains by comparing them to early Macintosh and IBM computers: "These different information-processing machines use the same wires, and when you enter the same data into their programs you get the same results—but the wires are organized differently and you must use programs designed for their differently configured systems."[20] That's why Pepperberg always says that Alex's cognition is "analogous" to ours, never homologous.[21]

Teaching Alex to Communicate

Before Pepperberg started her research, scientists had failed in two-way communication attempts with birds. Pepperberg did not fail, and for the reasons that others have succeeded in communicating with apes and marine mammals. She argues that a pupil of any species will learn a "communication code" if she's exposed to the meaning of its elements and the rules for combining them, and understands how the code is used and affects others.[22] When teaching communication codes, the closer one ties the teaching process to the student's natural skills, the easier the code will be to teach.

Communication is a social skill. Humans won't readily learn to communicate unless they are in social situations.[23] That's why such extremely socially deprived "wolf-children" as Kaspar Hauser and the Wild Boy of Aveyron never learned language. As wild parrots learn to communicate through social interaction, that's how Alex was taught. Because wild parrots naturally duet, Alex could learn by turn-taking. And he was educated in situations proven to help both mammals and birds learn.[24]

When trying to teach a bird a human-based communication code (or, presumably, teach a human a bird-based code), a teacher must honor four principles. Her pupil's level of competence must be respected. I didn't teach Christopher English by reading to him from *Finnegan's Wake,* but by speaking to him simply, often in "motherese." A student must understand how his learning relates to his life and why it benefits him. When a baby, if Christopher used a word to ask for something, he got it. When he made a connection between "dog" and "Marbury," I reinforced it. The most intense personal interactions are the most effective. Christopher didn't learn English by listening to me lecture to a group of babies; he learned by chatting with me, his mother, and his older sister, Roma. Finally, when "exceptional learning" is needed, the first three principles must be strictly applied.

Learning unlikely to happen normally is "exceptional." Certain kinds of learning may be exceptional because brains are wired to learn certain things only at critical times or at certain ages. Animals are normally wired to learn the communication code of their species easily, at least when young.[25] Alex could easily learn the communication codes of African Grey parrots. I would be hard-pressed to learn Chinese at age fifty. Christopher, at four, would soak it up, though human children with certain learning disabilities might not.[26] Learning the communication code of another species is exceptional. If Alex were to learn a human communication code, Christopher a parrot code, or me Chinese, the learning would be exceptional. This is why Pepperberg developed a modified "model-rival" technique in which teachers and students share the joint visual attention Pepperberg believes may be necessary for parrots, apes, even human children, to learn symbols.[27]

Here is how it works. Alex might perch on a chair and watch Pepperberg and an assistant chat about something in which he's interested, a banana, the color red, or the number five. He will even hear them intentionally make mistakes (5 percent of mistakes are unintentionally made by Alex's teachers; he usually repeats the correct answer and the human realizes his error).[28] Pepperberg often acts as teacher. Her assistant is both "model" for Alex and "rival" for the teacher's attention.

Their exchanges might appear spontaneous, but they're as stylized as Kabuki. After twenty years, Alex occasionally assumes the role of model-rival himself while helping to train other African Grey parrots in Pepperberg's lab.[29] But for many years, Pepperberg's laboratory functioned as a school of imitative learning for just one student.[30] Here's an example of a model-rival dialogue between Pepperberg and her assistant, Susan.

Pepperberg displays a banana piece.

"What's this, Susan?"

"Banana," Susan answers, with enthusiasm.

"That's right, Susan! Banana."

Pepperberg hands the fruit to Susan.

Susan might say, "Nut." If so, Pepperberg would roundly scold her, "No, Susan, banana!" and remove the banana from sight. Eventually she will bring it back and try again.

"What's this, Susan?"

"Nana," says Susan.

"You're close, say better!"

"Banana."

"That's right, Susan!"

And Susan gets the banana piece. Now the plot thickens. Susan turns to Alex.

"Alex, what's this?"

"No!" Alex says.

Susan looks at Pepperberg, "Irene, what's this?"

"Banana."

"That's right!" Susan says.

Holding the banana, Pepperberg turns to Alex. "Alex, what's this?"

"Nana," Alex says.

"Say better!"

"Banana."

"That's right!"

Pepperberg gives him the fruit. The object he labels is almost always Alex's reward; either that or he is offered something he likes better.[31]

African Greys can imitate words precisely. But they can imitate more than words. Bruce Moore studied the nonverbal imitation of one African Grey for five years. Several times a day he demonstrated unusual, often complicated actions for the parrot. He might open a door to leave the room with his right hand, wave with his left, say "Ciao," then close the door with his right. The parrot learned to do the same thing using his feet.[32]

Using her model-rival technique, Pepperberg treats African Grey parrots "as if" they were learning what she's teaching. Every successful language researcher does the same thing; only the failures don't. I did with Christopher and Siena what my parents did with me.[33] Sometimes Alex produces sounds resembling words he knows. Having learned "grey," he produced sounds similar to "grate," "grape," "grain," "chain," and "cane" when these objects were absent. Understanding that human language creativity can be ignited when adults respond to utterances as if they were intentional, Pepperberg responds "as if" Alex were intentionally referring to sound-alike words. This encourages him to use words creatively. When child psychologist Jerome Bruner visited Sue Savage-Rumbaugh's ape language laboratory, he found Kanzi treated not just as "if he" could learn language, but "as if he had human desires, beliefs, and intentions, and expects the same in return." To what end? In Bruner's opinion, "after several years Kanzi has become so like human beings that he finds his chimpanzee-raised sister's intersubjective insensitivity almost beyond forbearance."[34]

This "as if" idea has even infiltrated artificial intelligence. In 2000, Anne Foerst was a researcher at MIT's Artificial Intelligence Laboratory. She is a Lutheran minister, was the director of MIT's "God and Computers" project, and the theological adviser to scientists building two human-looking—and occasionally human-acting—robots, Kismet and Cog. We met these robots in *Rattling the Cage.* Each is built to trigger social responses in humans. Kismet has large blue eyes. Its eyebrows move. It has long lashes and, according to Foerst, a "cute, kissy mouth." She says the scientists there believe

> that humans are humans because we are social. Thus, we try to treat Cog and Kismet something like the way most of us treat babies *as if* they have intentionality, emotion, desires and intelligence. We give them as much social interaction as we can. . . . The point of reacting to Kismet is the same as reacting to a baby. We believe that only when you treat the machines *as if* they have all these social characteristics, will they ever get them.[35]

Alex's Imitation

In Chapter 4, we followed the development of Christopher's imitation as Piaget envisioned it. Now we see that Alex is a superb imitator. Why should we care?[36] Scientists used to think imitation was mindless mimicry. Now many argue it implies not mindlessness but the intelligent replication of another's behavior that has an end toward solving problems, and perhaps an ability to symbolize and mentally represent one's own mental states.[37] Some link imitation to self-awareness.[38] One animal copying another must be self-aware, they argue, to match the other's behavior.[39] Some think MSR (mirror self-recognition) requires imitation. Some believe imitation is a building block of theory of mind, others argue it's the other way around.[40] Recently discovered "mirror neurons" in the brains of macaques—neurons that fire not just when the monkey performs certain actions, but when she sees someone else do them—may play a role in imitation and theory of mind.[41]

Some say imitation is necessary for culture. Half a century ago, "culture" had at least 164 definitions.[42] At minimum, it involves a transmission of information that can affect another's behavior. In 1976, biologist

Richard Dawkins pronounced "meme" the unit of cultural evolution, as the "gene" is the unit of genetic evolution, claiming both operate by natural selection.[43] Language is a meme. So are singing "Happy Birthday," believing in God, piercing nipples, atheism, science, applauding school plays, and law. The idea that memes exist is a meme meme.

One way memes are passed on is by imitation. Susan Blackmore claims imitation is the only way. No imitation, no memes, no culture.[44] Only humans imitate, she says, therefore only humans have culture.[45] If she's right, imitation alone drives culture and that's a good reason to examine it. She defines imitation in two ways. One is very strict: the imitator must be able to decide what to imitate, make a "complex transformation" from another's point of view to hers, and match her actions to his.[46] Biologist Lee Alan Dugatkin complains how hard it is to prove that even humans pull that off.[47] Genes and memes, he says, are "replicators" able to produce many good and long-lasting copies.[48] But that doesn't mean they always produce perfect copies. Blackmore agrees. If "a friend tells you a story and you remember the gist and pass it on to someone else," she says, "that counts as imitation."[49] If that's imitation, Dugatkin ripostes, many species imitate, many have culture, and human and nonhuman memes must differ only in degree.[50]

Serious research into imitation began a hundred years ago. Psychologist Edward Thorndike thought imitation was "learning to do an act from seeing it done."[51] Many researchers still embrace his definition. But early on, scientists realized imitation could work on different levels, and higher levels might demand higher mental abilities. But agreement remains elusive on how many levels exist and what to call them.

Everyone agrees a low level of imitation can result from *social facilitation.* That means being around others who are doing something can make it more likely that I'll do it, like dancing at weddings. *Stimulus enhancement,* which may not produce imitation, occurs when an animal's attention is directed somewhere or to something by another animal. It probably accounts for the famous mid-twentieth-century episode in which birds began peeling back the foil caps from milk bottles in England. It was originally proposed that one bird had stumbled onto the secret and others had copied her.[52] A strong case was later made that the birds' attention had simply been drawn to the area in which there were milk bottles. Trial and error did the rest.[53]

The highest level of imitation has more than half a dozen synonyms: true imitation, reflective imitation, impersonation, action-level imitation, Stage 6 imitation, and more.[54] I'll call it *action-level imitation.* It seems to be what Alex does. *Emulation* happens when one animal watches another obtain food in some way; then, instead of exactly copying, the onlooker devises her own method to reach the food.[55] With respect to the ability to copy exactly, *program-level imitation,* says evolutionary psychologist Richard Byrne, lies between emulation and action-level imitation. But he argues that it's an intellectual cut above them both; it can never be the mindless mimicry that action-level imitation sometimes is because it demands an animal focus on the essence of a solution. One animal might copy neither the exact details of another's method to reach food nor her goal of obtaining food, but the overall structure of her actions, such as the sequence of stages.[56] Sometimes it's so hard to categorize imitation that comparative evolutionary psychologist Andrew Whiten thinks all the kinds of imitation lie on a continuum.[57] Whether they do or not, unlike action-level imitation, program-level imitation doesn't require an understanding that others have minds, just of the logical structure of the action sequences observed.[58]

Alex Mentally Represents

Joanna Burger gives an example of Tiko's ability to use insight. One day Tiko wanted Burger to preen his feathers. She was busy working at her desk and resisted his poking at her fingers and pulling at her pen. Suddenly he waddled to her bookcase and climbed to the shelf holding the glass in which she keeps his molted feathers. He pulled out and inspected several feathers, chose one, grasped it with his right foot, and began to scratch the right side of his head, neck, and face with the quill. When he was finished, he transferred the feather to his left foot and scratched the other side. Tiko had, Burger said, "conceptualized the problem and its solution."[59]

Pepperberg is sure Alex mentally represents the objects he labels and categorizes. With Griffin, Alex passed a standard series of fifteen increasingly difficult tests used to determine whether a human has achieved Piaget's full Substage 6 object permanence, plus an additional harder test researchers now believe more fully tests the concept.[60] The last three specifically test for full Substage 6 abilities. In one, an object is hidden within one container used to transfer the object invisibly to a second container. When shown the empty first container, will the parrots infer the object is in the second container?[61]

For the hardest task, Pepperberg played a trick. Previously she set out three plastic cups and openly used her hand to place an object under the first. With her hand she moved the hidden object under the second, then third, cup. The parrots passed the test if they either searched the last cup first or all three cups in the order through which Pepperberg had passed her hand. Now Pepperberg showed Alex an object and pretended to place it beneath the first cup. She really didn't. Then she moved her hand to the second, then third cup, continuing the charade.[62] Both parrots passed the test. Alex not only turned over the third cup, where the object should have been, but let out the "Yip!" he makes when startled.[63]

Alex expected to find something under the third cup, and his expectations were violated the way the expectations of infants are thought to be violated in expectancy-violation tests, such as the toy train and track I discussed in Chapter 4. But the test didn't necessarily prove that Alex had mentally represented the specific object Pepperberg had hidden, as opposed to some other interesting object he might imagine. So Pepperberg repeated the trick, this time showing Alex and Griffin she was hiding a very desired cashew. Through sleight of hand, she hid a less-desired standard nugget instead. When Alex and Griffin found the pellet, they acted as though their expectation—this time clearly their representation of the cashew—had been violated. The first time

Alex removed the correct cover and found the nugget, he turned away, slit his eyes, puffed his feathers, slightly opened his wings, and lowered his head. The second time, he turned away and began banging his beak on the table. Pepperberg, who would sooner eat glass than anthropomorphize, notes when Alex slits his eyes, puffs his feathers, opens his wings, and lowers his head, she makes sure she has a Band-Aid handy, for he is angry. And when he bangs his beak, he is either frustrated or displeased. Griffin threw all the cups onto the floor.[64]

Pepperberg assigned the parrots a last test so complex that many scientists suspect it tests abilities more sophisticated than just object permanence. Again, she placed an object beneath a cover. While one parrot looked on, she exchanged the position of that cover with one of two others. Both Alex and Griffin found the hidden object on the first try. Pepperberg concluded both parrots exhibit "full Stage 6 competence, putting them at the level of humans and great apes, and beyond many mammals and nonhuman primates."[65]

Pepperberg also thinks Alex mentally represents when he labels abstract symbols, generalizes his labels to new situations, and understands them to be interchangeable and recombinable in novel ways.[66] This is an extremely sophisticated ability. In Amboseli National Park in Kenya, I listened to vervet monkeys make different calls when they sensed different dangers. Each call was a "referential communication" because it elicited a different escape response from other vervet monkeys.[67] Alex's ability is even more complicated. He conceptually uses an abstract symbol to indicate the presence of an object in a specific situation.

After two years, Alex learned English labels for nine objects (paper, key, wood, hide [rawhide], peg wood [clothespins], cork, corn, nut, and pasta [bowtie macaroni]), three colors (rose [red], green, and blue), three-corner (triangle) and four-corner (square), and "nuh" (no).[68] Once he learned a symbol, say, "key," he used it for all sizes, shapes, and colors of keys, not just the specific key on which he was trained. Recall that after Alex learned "grey" by asking the color of his reflection, he produced "grate," "grape," "grain," "chain," and "cane," even when these objects were not present, not knowing what they were. But Pepperberg and her assistants placed a nutmeg grater, a grape, a parakeet treat made from grain, a chain of paper clips, and a piece of sugarcane before him to help teach him the actual words. As Alex began to understand that these sounds could be used as abstract symbols that stood for objects, he demonstrated his ability to mentally represent the object each word symbolized.[69]

Alex was ready for Pepperberg to ask a harder question. She knew he could mentally represent. He could label a key "green." But did he understand there exists a category called "color," of which "green" is one? If Pepperberg showed him a green anything, however shaped, whatever size, of any use, made from any material, would he know it was green?[70] Alex's mistakes while learning labels—he might err by labeling a clothespin ("peg wood") as just "wood" or a green key as "key"—hinted he was beginning to acquire an idea of category. [71]

Here's an excerpt from a model-rival training session on color and shape with Alex. Pepperberg is the principal trainer. An assistant, Kimberly Goodrich, is the secondary trainer.

Irene: (holding up a green triangular piece of wood) "Kim, what color?"

Kimberly: "Green three-corner wood."

Irene: (removes wood from sight and turns body slightly away from Kim) "No! Listen! I just want to know what *color!*" (turns back to Kim and shows her the wood) "What color?"

Kimberly: "Green wood."

Irene: (gives Kim the wood) "That's right, the color is *green;* green wood."

Kimberly: "Okay, Alex, now you tell me, what shape?"

Alex: "No."

Kimberly: "Okay, Irene, *you* tell me what shape."

Irene: "Three-corner wood."

Kimberly: "That's right, you listened! The shape is three-corner; it's *three-corner* wood." (gives Irene the wood)

Irene: "Alex, here's your chance. What color?"

Alex: "Wood."

Irene: That's right, wood; what *color* wood?"

Alex: "Green wood."

Irene: "Good parrot! Here you go." (gives Alex the wood) "The *color* is *green.*"[72]

He had learned "green" as a color. He had learned "three-corner" as a category of shape. Able to categorize the same object with respect to color or shape, he had learned to "reclassify." He could view the same object in multiple ways, decode a question presented in symbols (English) asking him to identify which of two categories of abstract attributes (color and shape) he was being asked and answer correctly.[73]

Could Alex now learn that objects could be the same or different? This is a more complicated task than the one the honeybees performed when they matched abstract symbols for a nectar reward. It involves a sophisticated ability to use the arbitrary symbols "same" and "different" to represent a *relationship* between objects. Could he understand that a blue key and a blue clothespin share the same color, that green paper and green cork do too?[74] When shown two novel objects differing in color, shape, and material, and asked "What's same?" and "What's different?" could he respond, not with the actual similarities or differences (red, triangle, wood), but with the correct *categories* (color, shape, material)?[75] He could.[76]

Could Alex understand nonexistence? This may seem trifling, but it is a deceptively complex mental ability. An animal responds to the absence of something when he *expects* something to happen that doesn't.[77] Alex has expectations. Recall his reaction during the object permanence tests. When the cashew is not where he expects it to be, Pepperberg had better look sharp. Alex's ability to respond to absence depended upon his expectation that something would be present.[78] Using the same techniques used to teach "same" and "different," Pepperberg taught Alex the concept of "absence." In the end, when shown a pair of objects, he could state whether they were same or different in color, shape, and material; when there were no differences, he could correctly answer "none."[79]

Harder still is the ability to relate abstract properties. Could Alex understand a tennis ball is larger than a golf ball, but smaller than a basketball? Pepperberg uses "matter" or "item" to stand for an object. When asked "What matter bigger?" and "What matter smaller?" he will tell you.[80] And he answers "none" when presented with objects of the same size.[81]

Could Alex verbally choose a correct answer from among multiple possibilities?[82] Pepperberg trained him to look at several objects of varying colors, shapes, and materials and answer such complicated questions as "What color is the item that is circular and rawhide?"[83] To respond correctly, he had to possess representational knowledge sufficient to allow him to understand each of three abstract categories, color, shape, and material. And he had to realize that he was supposed to combine them to answer the question. He succeeded.[84]

Alex Means What He Says

Before he acquired "want," Alex sometimes referred to "nut" when there was none about. Was he commenting upon a nut? Or was he requesting one? Pepperberg couldn't know, other than through an often ambiguous context. Alex would have to learn to request.[85] She taught him "want," figuring that if he could learn to use it properly, a window into his desires and intentions might open.[86] Requesting meant he would have to discriminate among many symbols for objects, then communicate what he wanted and what he didn't.[87] Teaching him to discriminate would be no problem.

She taught him to communicate what he did and didn't want with the model-rival technique. A trainer might hold up paper and cork. The model-rival would say "I *want* cork!" stressing "*want.*" The trainer would fork over the cork. Then they would reverse roles and repeat with paper: "I *want* paper!" Sometimes the model-rival simply responded "cork" or "paper." The trainer would say "Yes, this is a cork (or paper). Do you *want* it?" The model-rival would turn to Alex and say

"Alex, do you *want* cork?" If Alex responded "Cork," the trainer would say "I *want* cork."[88]

When Alex began to show glimmers of understanding "want," Pepperberg began to converse about "want." Here's one conversation between Alex and Pepperberg's assistant, Denise Dickson.[89]

Denise: "Well, what do you want, Alex?"

Alex: "Tea."

Denise: "Well, we don't have any tea. How about some pop? Would you like pop? Look, is this what you really want?"

Alex: "Want a cracker."

Denise: "Oh, you can have some cracker . . . good boy. (pause) Now look here . . . what's this? Ooooh . . . look."

Alex: "Want banana."

Denise: "Oh, good boy . . . have some . . . oh, good boy . . . you can have some of that banana. You didn't eat much . . . does it not taste very good? Do you want (it)?"

Alex: "Want some tea."

Denise: "Well, we don't have any."

Alex uses "I want," "wanna," and "want" appropriately and interchangeably to request interesting items.[90] He is fed whether he requests food or not. But if he wants fewer nuggets and more fruits and nuts, he must request them. And he does. When I visited Pepperberg's lab, Alex asked for nuts half a dozen times. Griffin asked for corn and a shower.

This use of "want" may reflect Alex's emotional states. But it implies some understanding of personal pronouns. His statement "Wanna nut" implies "*I* wanna nut." When Pepperberg places a favorite nut beneath a cup too heavy for him and Alex tells Pepperberg "Go pick up cup," this implies "*You* go pick up cup." When he asks someone to "Come here," he implies "*You* come here."[91] He can be explicit. "You tickle?" he asks, and "I want nut."[92] When Pepperberg left him at the veterinarian for lung surgery, Alex demonstrated both implicit and explicit understandings of "I" when he called to her as she was leaving "Come here. I love you. I'm sorry. I want to go back."[93]

Without training, Alex picked up the phrase "Wanna go" from his trainers, who would often ask him where he wanted to go. If he said "Wanna go gym" when he was in his cage, he was taken to the gym (a wood and rope structure). If he was already on the gym, Pepperberg would respond "Well, you're already on the gym."[94] If he said "Wanna go gym" and hopped onto a trainer's hand, but was carried to another location, he usually said "No," refused to hop off, and repeated, "Wanna go gym."[95] Lacking fingers and hands, he can't easily point toward the gym. But he may stretch his entire body in the general direction of where he asks to go and then sit quietly once placed there.[96] Significantly, he rarely, if ever, says something like "I wanna go nut" or "I wanna gym."[97]

Pepperberg tests his intentional use of words by watching his reaction when his requests are honored, and when they are not. If Alex requests an object and Pepperberg brings him something else, he usually refuses it, repeats his request, then eats or manipulates the requested object when it is finally produced. She conducted a formal test over several months. Alex would request an item. Pepperberg would present another. Eighty-six percent of the time, Alex responded "No." Sixty-eight percent of the time, after he said "No," he also repeated his request.[98]

Pepperberg points to a possible example of Alex's intentionally creating a word. When she began teaching him "apple," he knew "banana," "cherry," and "grape." At the end of the fall apple season, Alex stopped eating apples and Pepperberg stopped teaching him "apple." With spring, she began anew. A few days later, Alex looked at an apple and said, "Banerry . . . I want banerry," and bit the apple. Since then he has used "banerry" for "apple." Pepperberg thinks it might be a combination of "banana" and "cherry."[99]

Alex asks questions. "What's that?" he inquired about the cups used to test object permanence. He asked what a carrot was. And an orange.[100] Peering into a mirror, he asked Pepperberg's assistant, "What color?" The assistant replied, "That's grey; you're a grey parrot." After asking the same question six times, Alex learned "grey."[101]

Sometimes Alex looks at Pepperberg and says "I'm gonna go away." Then he does.[102] At other times he retaliates by bollixing her tests. Pepperberg says that when Alex is presented with an array of seven objects and asked a question, he has chosen the wrong answer twelve times straight. Because the odds of Alex's randomly erring twelve times straight are less than 1 percent, he's probably doing it on purpose.[103] Alex does this so often that an MIT assistant confessed to me her intense frustration with this behavior. She appeared so exasperated I asked her how long she had been working with him. "One month," she said. In Donald Griffin's view, Alex "gives every evidence of meaning what he says."[104]

Alex's Counting

Alex distinguishes quantities and has acquired the abstract concept of number.[105] But does he count? Counting demands he label objects with numbers as he might label colors or shapes or materials.[106] Pepperberg tested him by piggybacking atop these sophisticated abilities. On a tray she placed one orange chalk, two orange pieces of wood, four pieces of purple wood, and five purple chalks, and asked, "How many purple wood?" "Four," Alex answered. She displayed one blue box, three green boxes, four blue cups, and six green cups, and asked, "How many green cups?" Alex answered, "Six." And so on.[107]

Pepperberg always interprets her data conservatively. Alex's responses were, she said, consistent with being able to count. But it could have been consistent with quickly estimating number, called subitizing. If so, his ability to subitize is far greater than that of humans, Pepperberg says.[108] Maybe it is. She concludes, at minimum, he can "vocally label the quantity of a specific subset of objects in a complex heterogenous array . . . [and] that the 'take home-message' is that we should not underestimate the numerical abilities of nonhumans, particularly nonmammals."[109] She's still trying to nail down whether Alex can count.

Does Alex Have a Self?

In her third decade of work with Alex, Pepperberg thinks he thinks and thinks she's learned something about what he thinks about. She's willing to stack what she can prove against what others can prove about the mental abilities of highly intelligent, autonomous apes and marine mammals.[110] Grasping abstractions of labels and categories, of same/different, absence, conjunctions, relative size, object permanence, and acting intentionally, Alex compares favorably.[111]

African Greys are not unique birds. Magpies show Substage 5 object permanence and pass many Substage 6 tasks, though it's unclear whether they can pass them all.[112] An Illiger mini macaw and a cockatiel have passed them all.[113] Individuals of other bird species engage naturally in behavior that would probably allow them to do the same.[114]

Joanna Burger finds Tiko's memory "extraordinary" because he looks for objects in places from which they were moved years before.[115] Scrub jays appear to demonstrate an episodic-like memory—specific memories of experiences that, in humans, are associated with self-consciousness—in recalling where they cached food.[116] Some birds remember what they stashed, where they stashed it, and when; they will not bother to retrieve perishable food past the time it loses its freshness.[117] New Caledonia crows manufacture tools from twigs and leaves in a highly standardized manner, including hooks, to hunt small invertebrates, skills not appearing in humans until the Lower Paleolithic.[118] Moreover, they display a laterality in how they make their tools, generally moving from left to right, the only nonhumans known to demonstrate laterality in tool manufacture.[119]

Within the enormous amount of scientific literature on ravens are wonderful reports telling how ravens slide on their bellies in the snow, defend their nests with rocks, barrel-roll when alone, fly upside down, and yell in a functionally referential way.[120] After watching them for seventeen years, Bernd Heinrich concludes they rival African Greys in intelligence.

Ravens are not in the same order. African Greys are Psittaciformes; ravens are Passeriformes, perching birds. Heinrich has no doubt ravens are moody and feel anger, fear, love, and hate.[121] They form intense attachments with humans who embrace them as friends and family.[122] The reason is mutual communication. A trusting raven, Heinrich says, is "expressive, communicates emotions, intentions, and expectations, and acts as though it understands you."[123] He found raven dialects and behavior differ from place to place, evidence, he says, of culture.[124]

Because he works outside, Heinrich can't easily devise the elaborate tests Pepperberg gives to Alex. But he has made startling observations. Ravens routinely watch other ravens cache food, then rob the caches. For this reason, when being watched, some ravens will carry their food away to cache, which Heinrich believes is an attempt to outwit competitors.[125]

Perhaps his most significant observation occurred when he set out a large piece of suet in his backyard. Because it was frozen, it could be eaten easily only by chipping away small pieces. One unsatisfied raven concocted a plan to obtain a much larger piece. Aiming dozens of pecks in rows and ignoring the tiny flakes that fell away as he worked, he grooved a canal in the suet three inches long and half an inch deep. When the work was completed, a large piece would fall away. Heinrich startled the raven before he could complete the task, but the suet showed evidence of another successful groove. He has since seen similar grooves carved by other ravens.[126]

Heinrich decided to test ravens' insight: could they mentally evaluate choices, make intelligent decisions, and engage in intelligent behavior without engaging in trial and error? Heinrich needed to think up a problem that ravens could physically solve, but one so unusual and outside normal raven experience that they could not have learned the solution nor could they have been hardwired to solve it. He devised five studies centering around meat dangling from a thirty-inch string tied to a branch.[127] Ravens could reach the food only by perching above the string, reaching down, grabbing it with their bills, pulling up a length of string, stepping and pressing down hard enough to anchor it, releasing the string, then repeating the entire process until they reached the meat.

Many ravens succeeded, some in as short a time as thirty seconds, often within a few minutes. Ravens who failed sometimes attempted to fly away with meat that either Heinrich or another raven had pulled up, only to have it yanked away in a few flaps. But the string-pullers never did this, not once. This suggests that in their minds, string connected perch to meat.[128] Heinrich concluded that ravens can use mental images, insight, and intelligence to solve problems.[129] This in a creature with a microscopic cortex and a 15 cc brain.[130]

What biologist Douglas Chadwick, writing for *National Geographic,* said about Heinrich and ravens also goes for Pepperberg and Alex. Heinrich's

> study of mental processes in ravens was rejected by four academic publications before it finally appeared in *The Auk,* the journal of the American Ornithologist's Union. Science seems to be having enough trouble interpreting recent evidence suggesting that our closest relatives, chimpanzees and other great apes, possess reasoning abilities. The possibility that some raucous featherhead might have them too threatens to upend centuries of grand assumptions about humanity.[131]

Does Alex Have Practical Autonomy?

According to Joanna Burger, "[N]o one who has lived with a parrot will for a second doubt that they have thoughts and feeling similar to ours."[132] As I noted, Burger thinks Tiko's intelligence resembles that of a precocious three-year-old.[133] Pepperberg says Alex's cognitive abilities compare to those of a four- or five-year-old child, and he reasons, comprehends, and calculates at the level of a four-year-old. He intentionally communicates. Pepperberg says that in his ability to act intentionally, grasp abstractions of labels and categories, same/different, absence, conjunctions, relative size, and object permanence, Alex compares favorably to apes and marine mammals who, we will see, are advanced.

His use of "I," "want," "wanna," and other words in contexts that make it seem clear he understands who "I" is and distinguishes "I" from "you" strongly suggests self-awareness. His sophisticated mental abilities, including rousing success at Substage 6 object permanence tests, strongly suggest he has ecological, interpersonal, and conceptual selves. We don't know whether he has a temporally extended self. He may. Alex imitates. To the extent that involves self-awareness, the intelligent replication of another's behavior with an end toward solving problems, and an ability to symbolize and mentally represent one's own mental states, he possesses those abilities. To the degree it's a building block of MSR, he might be able to do that, too, although Bernd Heinrich agrees that no bird has demonstrated self-recognition.[134] Alex certainly mentally represents, understand symbols, uses a sophisticated languagelike communication system, and solves complex problems.

But for the following, I would place Alex in Category One. His evolutionary distance from humans is so great that even Pepperberg believes Alex's cognitive abilities are analogous, not necessarily homologous, to ours. We have no idea whether he has any elements of a theory of mind, for he has never been tested. And we don't know whether he would pass an MSR test because he has never been given one. We can, with confidence, place him in Category Two. I assign an autonomy value of 0.78. This entitles him to dignity-rights under a moderate reading of the precautionary principle.

Unlike the case of honeybees, no unusual physical or economic barriers prevent judges from assigning Alex the autonomy value that is an African Grey parrot's due. Unusual political barriers in the form of organizations of parrot owners might resist attempts to grant their parrots rights against them. There may also be psychological barriers to protecting a bird that might not exist for a mammal.

SEVEN

Marbury

We Are Family

Marbury is snoozing by my left foot. I inch my big toe toward him and gently stroke his muzzle. His right eye pops open. He raises his head and then, to my horror, begins to struggle to his feet.

"No!" I cry.

I hadn't intended that. He ignores me, gains his feet, and sets his chin on my left leg. We lock eyes while I stroke his shaven, stubbling, right shank. Four weeks before, he had cleanly severed a ligament and partially torn another just by jumping into a car. Three thousand dollars and two surgeons later, he is mending. Meanwhile, he's not supposed to be jumping around.

No one knows when humans and dogs began to share home and companionship. Fossils show we were living together 14,000 years ago. Genetic analysis of the mitochondrial DNA of dogs suggests domestication may have occurred more than 100,000 years ago.[1] If that's correct, humans and dogs have lived together about as long as there have been humans and dogs. Some argue our lengthy and intense social relationship led us to coevolve.[2] Some even see dogs as "man-made."[3]

I don't know Marbury's breed. He looks like a German shepherd with something else mixed in. I had no idea what that something was until my law partner and I were discussing strategies to save the lives of two condemned dogs. I asked to see their photographs. She flipped one across the table. I picked it up and started laughing. It was a picture of Marbury. But it wasn't. It was one of the condemned dogs, a Rhodesian ridgeback. Marbury lacks a ridge; otherwise, he's a ringer.

Rhodesian ridgeback, Pekinese, St. Bernard, miniature poodle, German shepherd, Chihuahua, it doesn't matter; they're the same species, *Canus familiaris,* of the family *Canidae,* order *Carnivora,* class *Mammalia.* All descend from wolves, whether one line of wolf or many, nobody knows.[4] When I asked my evolutionary biologist, "Mr. Darwin," for his estimate of when Marbury and I last shared a common ancestor, he said between 65 and 145 million years ago. His best guess was 90 million.[5] Hold your breath for thirty seconds. It's beginning not to seem so long ago.

My relationship with Marbury is so common that the human family has been said to be a dog's natural and social environment. Sleeping at my feet, Marbury is in his wild. More Americans share their lives with companion animals than with children.[6] We may enculturate them as we do our children, and our relationship is in some sense similar to that of parent and child.[7] Dogs have become extremely attached, even dependent, upon us, as our children are, sensitive to our desires and feelings, because for millennia we bred them to be that way.[8] "Can anyone doubt that the love of man has become instinctive in the dog?" Darwin (the real one) asked.[9]

Because we have lived together for so long, dogs who adapt best to a human environment appear to gain a selective evolutionary advantage.[10] Studies comparing the problem-solving abilities of dogs and tamed wolves may reveal one consequence of this selection. Wolves tend to solve trial-and-error problems better than dogs, and they aren't as amenable to being trained.[11] It may turn out that domestication made dogs worse trial-and-error problem solvers. But because the more independent working dogs solve problems better than companion dogs, the answer may be that we made them socially dependent; the more dependent a dog is, the worse she may be as a problem solver in strange situations, just as humans are.[12] On the other hand, when researchers cued seven dogs and seven wolves to which container of food was hidden by tapping on the correct container, pointing to it while gazing at it, and pointing to it while not gazing at it, all seven dogs used the cues to figure out where the food was. None of the wolves could.[13] Brian Hare, one of the experimenters, concludes that "dogs have clearly been selected to completely tune in to information provided by humans."[14] Perhaps philosopher Daniel Dennett hit the mark when he wrote, "By treating them as if they were human, we actually succeed in making them more human than they otherwise would be."[15] Unconsciously or not, we have bred dogs to our specifications and enculturated them. This chapter is largely about the abilities of dogs as we created them.

Amazing Dog Stories

Tick-infested Argos, half-dead on a dung heap, was the only creature on Ithaca to recognize disguised Odysseus after his two-decade absence. As Argos weakly thumps his tail, drops his ears in recognition, and dies, Odysseus flicks away a tear.[16] The Greek Stoic Chryssipus reports a hunting dog pursuing prey encountering a three-pronged crossroads. The dog sniffs the first road. He sniffs the second, then takes the third, without sniffing.[17] Applying formidable anthropological talents, Elizabeth Marshall Thomas trails her canine family through the streets of Cambridge, Massachusetts, and concludes that they fall in love and "marry," that romantic love is not just for humans but that "(f)ully as any human love story, the story of Misha and Maria shows the evolutionary value of romantic love."[18] Primatologist Roger Fouts relates when he was four, his beloved dog Brownie dives beneath the tire of the family pickup truck as it is about to strike Fouts's nine-year-old brother Ed. Not a family member "doubted for a second that Brownie had sacrificed her own life to save my brother's."[19] Psychoanalyst Jeffrey Masson writes an entire book fueled by anecdotes about canine emotions.[20] Plant physiologist Rupert Sheldrake argues that the human/dog bond opens "a channel for telepathic communication."[21] Michael W. Fox, who holds doctorates in veterinary medicine, human medicine, and ethology, thinks dogs, humans, and other social species are joined in an "empathosphere" allowing for communication by extrasensory perception with those with whom they have emotional connections.[22]

Every human companion, every dog trainer, thinks she knows how dogs tick emotionally and intellectually. Sociology professor Clinton Sanders has spent a decade exploring the human/dog relationship. He's learned "most clearly that humans see their dog companions as unique individuals who are both thoughtful and emotional."[23] "We understand their emotions," he says, "and draw from our experiences with them the knowledge that they can, in turn, feel something of what we feel."[24] Stanley Coren, psychology professor and dog trainer, writes that obedience trainers have "an implicit theory of dog intelligence" upon which they base their techniques.[25] However, trainers have disagreements. Some believe dogs think like young human children; others, that they use some logic. Or that they can learn only associations.[26]

I have puzzled over the minds of dogs personally and professionally for decades. Since I was three years old, I have lived with four dogs. One, I swear, understood me as well as I have ever been understood. Over twenty years, I have litigated 150 cases in which a public official ordered a companion dog killed or exiled (which generally leads to the same end) because of a "vicious disposition."

The trials have often revolved around a dog's temperament, character, and states of mind. I encourage officials to understand that what appears to be vicious behavior to them may be something altogether different from a dog's point of view. Feelings never run higher than when a child is bitten on the face. Parents demand the dog's death and officials obligingly order it. But dogs often gently warn other dogs away by nipping them on the muzzle. So they nip a child's muzzle. Except that a child doesn't have a muzzle and her delicate face tears much more easily than does a muzzle.

After studying dogs for thirty years, Dorit Feddersen-Petersen finds "a deficiency of quantitive data which is more than counterbalanced by an excess of unproven speculations."[27] Other researchers aren't so sure. Biologist Marc Bekoff could scarcely contain his scorn in a recent review of journalist Stephen Budiansky's book, *The Truth About Dogs,* chastising him for his "failure to recognize that critical studies have not yet been performed."[28] "Dogs are complex beings whose psyches and behavior are not easily understood nor teased apart," Bekoff wrote. "No one . . . really knows the truth about dogs."[29] Let's see what we do know.

A Potpourri of Mental Possibilities

Surprisingly little formal research has been done on the mental abilities of dogs, only a small fraction of what has been done on honeybees. Brian Hare thinks it's because dogs have been seen as "artificial," their traits selected by humans.[30] Detailed studies of communications between dogs and between dog and human are finally under way. Meanwhile, the high points of what we may know about canine selves, cognitive maps, dreaming, object permanence, memory, and insight are easy to summarize.

Dogs have fizzled a few MSR (mirror self-recognition) tests.[31] But, like elephants, they hear sounds above the human threshold. Famous for their sense of smell, they have five of the nasal turbinals specialized for odor detection, and 220 million scent receptors, compared to our paltry 5 million.[32] In some ways they live in another sensory world, hearing things and smelling things we can't, unable to sense things we can. This might affect their performance on standard MSR tests, but whether they lack self-awareness remains to be seen.

What of dog *selves*? Antonio Damasio believes dogs have autobiographical selves.[33] Frans de Waal and colleagues find it hard to imagine a dog who marks his territory with urine or finds his way home can do either without a self that allows him to understand where he is in the world. They conclude some self is necessary "to explain and answer most animal behavior."[34] It is likely dogs have an ecological and interpersonal self, as well as James's "I-Self," and probably parts of a "Me-self" as well.

Evidence is converging from several directions that dogs mentally represent. They probably carry cognitive maps in their brains of the sort that honeybees have, internal representations of the spatial relationships between external sites. We know they take shortcuts that require them to perform mental calculations.[35]

Does Marbury *dream*? He runs and sometimes barks and whines in his sleep. Other sleeping dogs appear to feed, bite, even copulate.[36] Hints he might dream recently emerged from a study of rats implanted with microelectrodes that recorded firings from neurons in the hippocampus, a brain structure important to forming and encoding memories. The rats were trained to run on a circular track. As they ran, a record was automatically made of the sequence of neuron firings; when the rats slept, the firing almost exactly replicated the pattern created when they were running on the circular track.[37]

In 1981, Estella Triana and Robert Pasnak reported that two dogs achieved Piaget's Substage 6 object permanence.[38] Today, ethologists describe object permanence for dogs as Piaget described it for human infants. If Triana and Pasnak's claim is true, dogs can mentally represent and are not chained to the here and now of the world of their immediate perceptions. They convinced Sylvain Gagnon and François Doré, who work with object permanence in nonhuman animals, that dogs at least attain Piaget's Substage 5. But Gagnon and Doré suspected flaws in the methodology that had been used to prove Substage 6.[39] The Substage 5 and Substage 6 tests they decided to adminster to thirty dogs, mostly terriers, were based on those usually given to human infants.

To pass the Substage 5 "visible displacement" problem, dogs had to track a rubber toy placed first behind one screen, then another, while they were watching. They succeeded by age seven weeks.[40] For the Substage 6 "invisible displacement" test, Gagnon and Doré put a rubber toy into a three-sided container and placed it behind a screen.

Then they secretly removed the toy and rotated the empty container so the dogs could see it was empty. Any dog who searched for the toy behind the screen passed. And the dogs did. Gagnon and Doré concluded dogs could "mentally reconstruct the trajectory of an object they did not directly perceive," and this ability developed sometime between the ninth and thirty-fourth month.[41]

But the dogs had a tougher time solving Substage 6 "invisible displacements" than they had figuring out Substage 5 "visible displacements," and they weren't as accurate as the great apes.[42] Gagnon and Doré hypothesized that dogs and chimpanzees might have the same ability to represent invisible movements, but that dogs have a harder time keeping a complex mental representation in their working memory.[43] They posed dogs an "invisible transposition problem." In plain view, they placed an object behind a screen, then moved the screen and object together. At the same time, they moved a screen that had nothing behind it. The two screens were moved simultaneously three ways: they switched positions, the screen with the object was moved and its place was taken by the blank screen, and both screens were moved.[44] they reasoned that the location at which an object disappears behind a screen burns into a dog's working memory and stays there unless they see the object moved or it just can't be in the same place anymore. Either alternative allows them to choose correctly.[45] The dogs were moderately successful; but their skills fell short of those of a twenty-month human infant because they don't appear fully to understand invisible displacements.[46]

Not only does Gagnon and Doré's work represent the latest understanding of the ability of dogs to understand object permanence, but they are the only formal experiments on canine *memory.*[47] Dog memory is the subject of many stories.[48] I will add one. I have tried several court cases in which a dog broke from her enclosure and, oddly, raced past numerous pedestrians to pounce upon an eight- or ten-year-old boy. The parents always demanded the dog's death. But investigation revealed that the boy had been throwing rocks at the dog or hitting her with a stick over a fence, for months. She remembers.

No formal experiments have tested whether dogs demonstrate *insight,* the mental ability to "see" a relationship. But ethologist Lesley Rogers offers this anecdote: a blind dog was sent from England to her new home in Australia, where a flight of stairs led from up the front door. The dog learned to climb the stairs by running her nose across the width of each tread before stepping up. One day, she merely stood at the bottom of the stairs, then raced up the flight, head up; and that was how she climbed the stairs from that day, though she used her first technique to climb other flights.[49]

Dog Play

While Frans de Waal was being interviewed for the *New York Times,* a reporter's dog raced up, dropped a toy in his lap, and barked.

"Your pup is very clear about what he wants—he'd like me to play with him," de Waal said. "Sometimes I read about someone saying with great authority that animals have no intentions and no feelings, and I wonder, 'Doesn't this guy have a dog?'"[50]

Joanna Burger, the ornithologist owned by Tiko the parrot, echos de Waal: "I find myself amazed . . . that anyone could doubt that the animals closest to us—dogs, cats, horses, parrots (especially parrots)—have emotional responses to the things around them, or that anyone could question the proposition that they form ideas about the situations they find themselves in or the people they meet."[51]

If dogs do anything, they play. While walking your cocker spaniel, you might meet me walking Marbury. He may bounce over to "play-bite" your dog. Some scientists think play involves pretense, and pretense requires third-order intentionality. Does Marbury *want* your dog to *believe* (and maybe you) that he *wants* to bite her in play and not in aggression? Your cocker won't then *believe* she is about to be attacked. If intentionality is all-or-nothing, Marbury would have to understand every possible situation in which a third-order intentionality might arise.

As much as I adore Marbury, I don't think he can do that. In fact, he doesn't appear, in general, to have attained even a second-order intentionality. If intentionality is all-or-nothing, Marbury isn't "playing" with your cocker spaniel. It just appears that way. But Siena and Christopher, at age three, didn't appear, in general, to have attained a second-order intentionality. So they couldn't play. But they appeared to, and with Marbury. After a day with them, I was certainly pooped from doing *something*.

Almost twenty years ago, philosopher Ruth Millikan argued that intentionality need not be all-or-nothing. Some species might have evolved abilities to read only some mental states, or only the mental states of members of their own species, or only the mental states of a few species.[52] According to Bekoff,

> (w)hen most animals play, they use specific signals to signal their intentions—their desire to play and their beliefs that if they perform a particular behavior, then play will occur—and they also perform behavior patterns that are typically performed in other contexts, but whose meaning is changed in the context of play. The richness and diversity of play can only be captured by using mental terms.[53]

Marc Bekoff and colleagues argue that the rollicking, unstructured, uncontrolled, fun aspects of play demand a "richly developed cogntive system."[54] Bekoff concedes we can't be sure two playing dogs have beliefs, much less beliefs about beliefs. But Marbury, and maybe three-year-old Siena and Christopher, may possess a third-order belief detector specialized just for play, which may have been a lot easier to evolve than a general third-order belief detector.[55]

Canine Communication

In the days before Marbury's surgery, I would catch his eye and ask, "Wanna go out?"

He would jump up. Or he might stand at the front door and gaze at me, or bark. As we live in a semirural area, I would fling open the door, and out Marbury would bound, the screen banging shut behind him. Minutes later, he would stand outside the door and bark until I opened the door, and in he came.

In the mid–1990s, a team of Hungarian ethologists began a continuing set of experiments on canine cognition in general and what dogs understand about communication in particular. Vilmos Csányi argues the human-dog coevolution means not only are we interested in dog minds, but they are interested in ours and may have puzzled something about us out. This should not be surprising; remember, dogs have been selected, unconsciously or not, to be sensitive to humans, and we enculturate them.[56] Dogs may be expert at interpreting human cues because generation upon generation of human-attuned dogs have been bred for that purpose.[57]

Despite 100,000 years of companionship, we know surprisingly little about how dogs communicate; we're just starting to figure out what barking means. Some say it doesn't mean anything but excitement,[58] but research is beginning to contradict that claim. Dorit Feddersen-Petersen's preliminary sonographic analysis of dog barks is one of only two objective analyses of barking that exist, and it shows that different barks seem to correlate with different situations, such as playing, fighting, and threatening.[59] A second, yet unpublished study, led by Karen Overall, former director of the behavior clinic at the University of Pennsylvania School of Veterinary Medicine, agrees with Feddersen-Petersen. Different emotional situations, say, in a stressed or distressed dog versus a calm and unstressed dog, spur different barks.[60] Feddersen-Petersen also says that dogs have separate barks for elephants, humans, monkeys, strange cats, even individual barks for individual dogs.[61] Again, we're just beginning to understand barking.

Does Marbury know we're trying to communicate with him? Or does he just associate my look, tone of voice, and sounds with the fact that if he jumps for the door, I'll let him out?[62] We understand human-to-dog communication better than dog-to-human or dog-to-dog. Stanley Coren gives us a "minidictionary" of his dogs' "vocabularies." It contains sixty-three words ranging from "Away: the dog moves back from whatever it was investigating or attending to" to "X-pen: this causes the dog to wait near the exercise pen until I place it inside."[63] Mix in twenty-five gestures for "sit," "stay," "heel," and the like, which he equates to sign language, and the "total receptive vocabulary" of his dogs is ninety.[64]

The Hungarian team is more cautious, acknowledging "dogs seem to be able to extract information from complex social situations that occur during interactions with their human companions."[65] But even when they asked thirty-seven humans to answer a questionnaire about what words they believe their dogs understand, they took the responses as reflecting the humans' opinions, not the reality, of what the dogs actually understand.[66] Those opinions were that the dogs understand at least short sentences reliably and well.[67]

The Hungarian team found mature dogs seem to understand about forty expressions. The individual variation is large, one dog comprehending just seven, another eighty. But they concede there is, as yet, no hard evidence for the exact role specific words play in causing dogs to respond.[68] No scientist claims that true language is involved. They're not sure what is. Csányi speculates dogs understand words not as symbols but as signals.[69] Even Coren detects no canine ability to understand grammar and thinks dogs "treat each string of sounds as a single, fixed, linguistic unit."[70]

Little investigation has been done. Early last century, two comparative psychologists, C. J. Warden and L. H. Warner, tested Fellow, whom they described as that "famous movie-actor dog."[71] Fellow's trainer claimed the dog knew more than 400 verbal commands.[72] After testing, the psychologists issued a press release stating he probably didn't understood words "as words," but had learned to associate sounds with commands.[73] However, Fellow was, they believed, a "most extraordinary dog," and if further tests revealed evidence of cognitive abilities that were "more than suggested" by their testing, "the dog would merit a much higher rank in the scale of mental evolution than most of us have been willing to accord him."[74]

Despite the psychologists' suggestion and the interest of tens of millions of humans in how smart dogs are and how well they communicate, it was almost sixty years before scientists delved deeper into the canine mind. It began slowly. In 1985, Patricia McConnell and Michael Bayliss examined fourteen shepherd/border collie herding teams to see how human shepherds communicate with their border collies. The subjects of the communications were primarily "fetch," "stop," "go left," and "go right," and were communicated through speaking and whistling. The dogs usually responded by acting more or less vigorously toward the sheep, but didn't appear to comb out any further information from the signals.[75]

Amazingly, as late as 1998, the Hungarian team could write that "apart from a few anecdotes in the literature on dog behaviour there are no studies examining the ability of dogs to respond to cues given by humans."[76] But serious and sustained research has begun. In 1997, the Hungarians set to work. They found dogs used all sorts of human directional gestures, pointing, bowing, nodding, head-turning, even glancing, to figure out which of two bowls contained food.[77]

Brian Hare and professors Michael Tomasello and Josep Call, all experienced in primate cognition, independently began their own series of human-dog communication studies. They tested whether Oreo, a twelve-year-old Labrador retriever trained to hunt, and Daisy, Hare's three-year-old mutt, could understand human pointing and gazing toward hidden doggie treats. Oreo and Daisy turned out to be "quite good," troubled only when Hare tried to fool them by looking one way while keeping his head straight.[78]

Hare also enticed them into playing fetch. He threw a ball. The dogs returned it. If they dropped the ball where he could see it, he threw it again. It turned out not to matter whether Hare was facing them or turned round, the dogs always dropped the ball where he could see it.[79] The scientists concluded two things. First, both dogs understood they were communicating with the front of a human, not his rear. Second, primate and dog behavior is here the same, except that some chimpanzees follow gazes alone, but dogs may not be able to do so.[80]

Hare and Tomasello tried to replicate and extend their findings with nine dogs and Daisy. For one trial, the investigators placed food behind one of two barriers. One human squatted beside the barrier that hid the food, then looked back and forth between the barrier and the dog. For eight of the ten dogs, that was enough to cue the food's location. In another experiment, a human stood equidistant between the two barriers, gazed at the one hiding the food, and pointed to it. That was enough to clue five of the eight.[81] In a third variation, one of the dogs, Maggie, stood next to the hidden food. Six of the ten dogs understood that she was telling them that this is where the food is. In the last test, Maggie stood equidistant between the barriers and gazed at the hidden food. Four of the six realized what she was doing.[82] Two dogs ended up using both cues, whether given by a human or dog. One was good with the standing or squatting cue, whether given by a human or dog, and two understood cues only when they were given by another dog.[83] Interestingly, the youngest dog, Uma, a six-month-old Labrador retriever, understood only Maggie's cues; the two oldest dogs, Daisy and Maggie, understood only human cues.[84]

In 1999, Tomasello, Hare, and another colleague, Brian Agnetta, wondered whether puppies could understand human cues and whether dogs of any age could interpret a physical object placed in front of hidden treats as a communicative signal telling them "Food is here!" Chimpanzees and orangutans fail at this; humans succeed only from the age of thirty months.[85] They found that at every age, the dogs were superb at locating hidden food when they saw a human indicate its location in some way, even by gazing toward it. "It is now very well established," they wrote, "that dogs have remarkable skills in the basic object choice task, much better than nonhuman primates." Of course, that doesn't mean we understand much about the psychology behind how dogs accomplish these tasks.[86] Because of the way they structured the tests, the scientists couldn't tell whether the dogs broadly understood human communication or only narrow cues indicating the direction of some object, or even whether their understanding of communication was related only to food-finding.[87] Hare writes that even the most parsimonious hypothesis is "very impressive," for it "shows dogs' ability to socially learn from another species *extremely* rapidly. If this was a human baby . . . there would be no hesitation to grant an understanding of intentionality."[88]

Meanwhile, the Hungarian team began looking into whether dogs "show." Can dogs intentionally communicate by directing someone's attention to something?[89] Had Marbury been a Hungarian dog and I a Hungarian dog companion, the researchers would have allowed Marbury to watch them place food or a toy into one out of three bowls, then raise that bowl high above his reach. I would have entered the room, sat and read for a minute, then given the food or toy to Marbury if he had indicated where the goodie was hidden. They would have compared this to what he did when alone with the food or toy or what he did when I walked in, petted him, read for a minute, and otherwise ignored him.

They found that had I been in the room paying attention to him, Marbury would have decreased the time he hung around the exit, increased his mouth-licking, sniffing, and looking at me, and started barking and alternating his gaze between me and the goodie bowl. They interpreted the barking and owner-gazing as attention-getting attempts. Gazing at the goodie bowl, especially when alternating looks to the human, was understood as a directional signal.[90] The scientists concluded that dogs are capable "of intentional, functionally referential communication," perhaps at a "level of understanding of the mental state of attention to that of the chimpanzee."[91] They cheerfully concede, however, that they cannot yet refute arguments that the dogs were conditioned in some unknown manner to respond as they did. But they think the likelihood is far less than the explanation that dogs have been selected for their ability to communicate with humans and, by golly, they can![92]

A year later, the Hungarian team replicated a study that had compared the ability of human three-year-olds and juvenile chimpanzees to interpret an experimenter's gaze as a mental state of attention.[93] The chimpanzees had flunked the test, though the Hungarian scientists thought the reason was probably that the children used in the experiment were enculturated and the chimpanzees weren't. Would dogs, like children, understand that when an experimenter turns her head and gazes toward a bowl, it contains food? Or that when she gazes at one bowl without turning her head, or turns her head and gazes above one bowl, it communicates nothing? They hypothesized that because dogs have been enculturated and selected to be sensitive to human gestures, they would perform in a way similar to that of children.[94] And they did, with one difference. The dogs acted as though gaze alone communicated something, for they avoided the bowl toward which the human had turned and was gazing, perhaps because human stares provoke submissive behavior in dogs.[95] As the original researchers had interpreted the actions of the children as meaning they understood the human gestures were communicating information, the Hungarians now interpreted the actions of the dogs in the same way.[96]

Years ago, scientists learned that dogs signal a desire to play to other dogs with play bows, barking, face pawing, leaping, and other ways that don't involve contact.[97] Humans like to play with their dogs. But when a team of British scientists filmed twenty-one human companions trying to induce their dogs to play, they found the humans often failed to understand what signals their dogs thought meant the humans wanted to play. When they used a signal that Fido understood to mean they wanted to play, he responded. But they used all sorts of signals, such as picking Fido up or pulling his tail, that he probably found threatening, not playful.[98] The scientists had humans test four play signals on Labrador retrievers kept in a laboratory for nutrition research. None of the signals involved physical contact. Sometimes they added vocalizations. Three of the play signals did indeed elicit a playful response in the dogs, especially when accompanied by vocalizations. The British team concluded that "domestic dogs have the capacity to interpret play signals given by humans."[99]

Even the role of guide dogs for the blind is being reexamined. Guide dogs were long believed to lead blind humans, as they had been trained. But the relationship between dog and blind human appears more complex. Human and dog cooperate and synchronize their movements. Both initiate actions and frequently exchange roles of leader and follower, and at two levels.[100] If the dog takes a specific action—start, stop, go around—the other follows, better to avoid the tangle of legs on a public street. And they cooperate to achieve larger goals, such as going to the store or visiting the library.[101]

Marbury Revisited

We end where we began, with Marbury. It's a warm late spring morning. I ask Marbury if he feels up to playing with the twins in the backyard. He immediately heads out the backdoor, barks, then hobbles about, mostly on three legs, more activity than would have made his surgeons happy. Finally I ask if he's ready to come in. He heads for the backdoor and upstairs for a drink of water.

"C'mon downstairs."

We sit before my computer, I to finish this chapter, he to fall asleep. His legs start moving rapidly and he's making muffled noises. I think he knew what I was saying in the yard. He is a very intelligent dog.

Does Marbury Have Practical Autonomy?

On present evidence, I place Marbury in Category Two. But how high? Higher than honeybees. The evolutionary distance between Marbury and me is nowhere as vast as the distance between honeybees and me. It is similar, we'll see, to that of elephants and whales, a seventh that of honeybees, a fourth that of Alex's. As we are both mammals, I can feel more confident that I have some better understanding of Marbury's take on the world.

There is virtually no evidence that honeybees have any sense of self. On the other hand, as we said before, Antonio Damasio thinks dogs have autobiographical selves, like humans and apes. If he's correct, Marbury should be placed at 0.80, perhaps even 0.90, and into Category One. But this still has to be proved. It is likely, though, he has an ecological and interpersonal self, as well as James's "I-self," and probably parts of his "Me-self." He can pass Substage 5 object permanence tests and perform moderately well in Substage 6 object permanence tests, though not as well as a twenty-month-old human. He probably carries cognitive maps in his brain and mentally represents. Anecdotal evidence suggests he insightfully solves problems. He vigorously plays, which may require substantial cognitive abilities. Not only does he desire and act intentionally, but in some circumstances, he understands human intentions, manifested by gazing, pointing, and verbally communicating, better than chimpanzees and orangutans, and at the level of human two-and-a-half-year-olds. He intentionally communicates.

On present evidence, I place him in Category Two, at an autonomy value of 0.68. Unlike Alex, he doesn't understand symbols, use a sophisticated language-like system, doesn't clearly exhibit Substage 6 object permanence abilities, or clearly suggest an ability for self-consciousness. This value emphasizes, however, that the serious study of canine mental abilities remains in its infancy. Early data are encouraging and point toward practical autonomy, but more work needs to be done. It won't take much to push Marbury to 0.70, perhaps beyond, and to entitlement to basic rights. Meanwhile, the possibility of his practical autonomy should be taken very seriously.

Economic and political obstacles may slow Marbury's gaining of dignity-rights in ways that would not slow Alex's. There are many more dogs than African Grey parrots, at least in the United States. More than 50 million dogs reside with more than one-third of Americans.[102] As with African Grey parrots, there are organized owners' clubs, but with considerably more clout. Upwards of 40 percent of dogs are registered with the American Kennel Club, the United Kennel Club, or similar organizations whose members are often obsessed with purity of breed and with competing in dog shows; they loathe the idea they might not be able to manipulate the bodies, even the genes, of dogs at any time and in any way they see fit. The thousands of commercial dog breeders, often tied into the kennel clubs and doing a couple hundred million dollars of business each year, are no more sympathetic.[103] We can expect only resistance from those who breed dogs for biomedical research, use dogs in biomedical research, or race them.

On the other hand, a hefty majority of Americans who live with dogs see them as family.[104] People in almost a third of 122 studied families reported they felt closer to their dogs than to any other family member.[105] American surveys reveal that about 90 percent of people are concerned about dogs and believe they are conscious, intelligent creatures able to reason and feel.[106] This is why in a case involving a dog one Texas Appeals Court judge urged that courts "not hesitate to acknowledge that a great number of people today treat their pets as family members. Indeed, for many people, pets are the *only* family members they have."[107] And that explains why libraries and bookstores are awash in books by veterinarians instructing how to keep Fifi healthy and treat her illnesses, by trainers telling how to make Fido do what you want, and by breeders singing the praises of one of the 140 or so varieties of dogs we have created. Alan Beck and Aaron Katcher, experts on the human/companion-animal bond, report that dogs are not "just any member of the family . . . they are children."[108] That there are so many dogs and that such a large percentage of people embrace them as family, even as children, may make it easier for judges to place them where they belong.

EIGHT

Phoenix and Ake

I have never met Phoenix and Ake. Louis Herman never permitted it. The controversial Herman runs the University of Hawaii's Kewalo Basin Marine Mammal Laboratory (KBMML) and its affiliated The Dolphin Institute (TDI), where the two dolphins are held. I sought his permission for two years. Between October 1999 and October 2001, I begged Herman six times to allow me to visit, and also to send me copies of the numerous unpublished masters' theses on dolphin cognition his graduate students have written over the years.[1] Herman ignored me until my fifth request in April 2001. I said I had completed a draft of this chapter and asked him to review it to ensure I had related his work accurately. To my delight, he responded. He and his colleagues love dolphins, he wrote, but

> [y]ou seem to wonder and to be sorely pained about my prior reluctance to communicate with you. I hope I am mistaken, but your prior communications to me suggested that you had a strong anti-captivity bias and consequent agenda to use my work to bolster your case, and in doing so, to misuse me. If so, how sad that you would use the knowledge we have gained through our laboratory studies to obviate any future knowledge we might gain. It is sad, also, that those who are so against captivity of the few scarcely raise a voice or take the necessary action to protect wild dolphins from the immense harassment and slaughter they suffer as a result of human activities, intentional or not. It must be reinforcing, I suppose, to pontificate, to possess a moral stance superior to

> others, and to see the world as dichotomous—good guys and bad guys, but no complex guys or complex issues. You will probably say—"but have you read my work." No, I have not, but have you seen me work? . . . I will read what you have offered and . . . protect myself against wrong statements.[2]

The next day, I sent Herman the draft chapter on Phoenix and Ake, and threw in drafts of the Koko and Alex chapters. Again, I begged leave to visit the dolphins. Six months later, his critique of this chapter arrived, but no permission to visit and no masters' theses. He further explained his reluctance to communicate with me. "This is a tough business I'm in, maintaining dolphins in a laboratory. No matter how much we care for and love our dolphins, we are too often beleaguered by extremists who have no respect for human rights while championing animal rights."[3]

Taxonomy and Brains

There are dozens of dolphin species, spinner dolphins, striped dolphins, spotted dolphins, white-sided dolphins, hourglass dolphins, and more.[4] Phoenix and Ake are Atlantic bottle-nosed dolphins, *Tursiops truncatus,* a toothed whale of the order *Cetacea,* suborder *Odontoceti,* family *Delphinidae.* But is there one species of bottle-nosed dolphin, or several?[5] Scientists disagree. So whenever possible, I will limit my examples to dolphins whom the scientists who observe them call *Tursiops truncatus.*

Mr. Darwin said I last shared a common ancestor with Phoenix and Ake the same time I shared one with Marbury and Echo, perhaps 90 million years ago. Sixty or 70 million years ago, Phoenix and Ake's land ancestor returned to the sea.[6] That event was pivotal to the history of dolphin brains, for that's when they began to develop differently from primate brains.[7] Beginning about 50 million years ago, primate brains began to grow more modular. But dolphin brains continued along the plan typical of the land brains they had left behind. Today their brains are built more along the lines of hedgehog and bat brains than human and other ape brains. Even the neurons of dolphin cortexes look and are organized differently from ours.[8] But how their brains grew!

Brains relate to cognition and mind, but no one is sure how. What, for example, do bigger or heavier brains signify? No one even knows how best to compare brains, though at least nine proposals exist. A couple are expressed in logarithms, which only mathematicians will love. Then there are absolute brain size, the ratio of brain size to body size, and the proportion of cerebral cortex to the whole brain. Each proposal captures an interesting aspect of brains. Size appears to matter, if not for the entire brain, then for critical portions of it. But no single measure tells us everything. And within a species, the smaller females average smaller and lighter brains.

Neuroscientist Susan Greenfield finds it "astonishing" that human brains quadruple (a 300 percent increase) between birth and maturity; the brains of other apes, and monkeys, increase only between 60 percent and 240 percent.[9] Dolphin brains increase between 200 percent and 240 percent.[10] This, she says, allows a staggering numbers of brain cells to connect in new and numerous ways.[11] But human brains are not unique.[12] If Greenfield finds a 300 percent human brain growth astonishing, what must she think of the growth of wild boar brains (480 percent), wolf brains (970 percent), fox brains (1,190 percent), or polar bear brains, which grow by an astonishing 4,510 percent?[13] Something else must explain human cognitive uniqueness. Perhaps it will turn out to be a combination of abundant brain growth coupled with a lengthy childhood.[14] Maybe it's the ways in which the frontal lobes are organized.[15] We don't know. But I suspect the differences will be complex and subtle.[16]

Cortex seems important because that's where information is received, organized, analyzed, and stored, at least in mammals.[17] Hanging over the eyeballs is the frontal lobe, the most forward part of its four lobes.[18] Its most forward part is the prefrontal association cortex. This part, probably interacting with others, is associated with agency, creativity, artistic expression, planning, decisionmaking, working memory, language, and other complex mental abilities.[19] Many used to think that the human frontal lobe was better developed and disproportionately larger than it is in other apes. But recent analysis suggests that the differences aren't large and that the frontal lobes of all great apes, and humans, are put together in similar ways. They count for just about the same percentage of brain, 36.7 percent in humans, 35.9 percent in chimpanzees, 35.4 percent in orangutans, and 32.4 percent in gorillas.[20] Orangutan brains average about 170 cubic centimeters when they are born and increase about 240 percent; their frontal lobes occupy about 35.3 percent of their brains, just shy of the value for chimpanzees.[21]

Gorilla brains average about 500 cubic centimeters in volume and weigh about 500 grams. The volume of a human brain is about 1,330 cubic centimeters and weighs about 1,500 grams. Chimpanzee brains have a volume of about 400 cubic centimeters and weigh about 400 grams. The brains of bonobos have a volume of about 340 cubic centimeters. Orangutan brains average about 335 cubic centimeters and weigh about 333 grams, somewhat smaller than a chimpanzee's and gorilla's, somewhat larger than a bonobo's. African elephant brains weigh about 7,500 grams.[22]

Dolphin brains weigh slightly more than human brains; the ratio of their brain weight to body weight exceeds that of the great apes.[23] The surface area of the dolphin cortex is enormous in relation to the rest of the brain, and compared to human brains (3,700 centimeters squared to 2,300 centimeters squared). Its surfaces are even more convoluted than ours. Because the dolphin's cortex is about half as thick as ours, however, the brain volume is about 85 percent of ours (560 cubic centimeters to 660 cubic centimeters), still greater than the volume of the chimpanzee brain.[24] Dolphin brains are asymmetrical, too; asymmetry in humans is associated with such sophisticated mental abilities as language.[25]

Dolphin Language Studies

In the 1970s, Louis Herman began teaching a simple acoustic language he invented to Keakiko, a female dolphin he nicknamed Kea, at KBMML.[26] He used another female, Puka, a veteran of unsuccessful language experiments with another scientist, for visual research.[27] Both dolphins had been longtime research subjects of the United States Navy; both had been captured, Kea eight years before Herman began his language work, Puka, twelve.[28] Herman had held Puka for eight years, Kea for five, in concrete seawater tanks measuring 15.2 meters (about 50 feet) in diameter and 1.7 meters (about 5 feet) deep.[29]

Five months after Herman began, Kenneth Le Vasseur and Steven Sipman, formerly Herman's students and participants in his behavioral studies with the dolphins, then primarily dolphin tank cleaners, concluded that requiring Puka and Kea to participate in experiments while confined to their tanks for the rest of their lives was morally and legally wrong.[30] Le Vasseur thought Puka highly stressed. He later testified she was erratically participating in Herman's experiments, often refusing, and engaging in neurotic jaw-snapping and tail-slapping, coughing for hours at night, and beating her head against experimental apparatus until she drew blood.[31] In the first hours of May 30, 1977, the two men plucked the dolphins from their tanks, placed them on stretchers, carried them to a padded van, and left a note identifying themselves as "The Undersea Railroad." After an hour's drive, they released Puka and Kea to the Pacific. Then they called a Honolulu press conference.

The police were unimpressed. Sipman and Le Vasseur were charged with taking more than 200 dollars worth of property from its owner with the intent of depriving the owner of it permanently. This was felony first-degree theft. Gavan Daws has written:

> In the view of Le Vasseur and Sipman, no more than humans should dolphins be forcibly captured, taken from their free state and impris-

> oned, as at Kewalo, for no crime, in isolation, unable to communicate or have physical contact with others of their kind, in featureless concrete tanks like water-filled cells, doing repetitive labor at the command of others who were in absolute control of their working hours and food supply. This was a life sentence against which there was no appeal, hard labor on rationed food in solitary confinement until death.[32]

Le Vasseur was tried first. Dolphins were persons, not property, he argued. The judge demurred. The dolphins had the right not to be enslaved under the Thirteenth Amendment to the United States Constitution. Not so, said the judge. The release was symbolic speech, akin to burning draft cards, protected by the First Amendment to the United States Constitution. The judge said no. By releasing the dolphins, Le Vasseur had committed the lesser crime of theft to prevent the greater crime of cruelty to animals. How so? asked the judge. Stealing the dolphins was a felony. Cruelty to animals was just a misdemeanor. Le Vasseur was convicted and sentenced to six months in jail and five years probation. The Hawaii Appeals Court affirmed everything. A more sympathetic trial judge would later revise Le Vasseur's sentence so that he served no jail time but was required to perform 400 hours of community service. Sipman's trial roughly followed suit.[33]

These events explained Herman's failure to cooperate with me, for he chastised me for "an insensitivity . . . to the trauma my students and I experienced when Kea and Puka were taken from us and without doubt (make no mistake about that) sent to their death. That trauma left an indelible scar, for those dolphins were loved as much as one would love one's own child. And, it doesn't matter that it was more than 20 years ago. The pain remains."[34] Perhaps his refusal to communicate is also connected to my representation of Rainbow and Kama in the early 1990s, when I repeatedly sued the United States Navy and other federal agencies, as well as the New England Aquarium, over the manner in which dolphins were being transferred. Herman has had many contacts with the United States Navy, has worked under contract for the navy, and used dolphins previously used or procured by the navy. Herman has written that Kea and Puka were "abducted," something that the *Oxford English Dictionary* implies happens only to persons. This is made explicit is its definition of "abductee," defined solely as "[a] person who has been abducted."[35] Had they been persons under the law, Kea's and Pukas's lives would have been very different.

What happened to Puka and Kea? Herman says they died.[36] He imagines they suffered extreme physical and emotional distress when released into the Pacific.[37] Ken Le Vasseur says eighteen months after their release, Puka was seen off Molokai swimming with a baby; Kea was spotted swimming with a pod of spinner dolphins near Kuai.[38] Steve Sipman lists more than a dozen purported sightings of one dolphin or the other over the next two years.[39]

Herman replaced Kea and Puka with another pair, two-year-olds captured off Gulfport, Mississippi, two kilometers apart. This was so close that Herman thinks they belonged to the same school.[40] He named one Phoenix, he said, to "symbolize the rebirth of the laboratory after the tragic loss of Kea and Puka." Akeakamai, or Ake, "a Hawaiian word meaning 'lover of wisdom' . . . symbolized the hopes for the future."[41] I asked Herman how, knowing what he knows about dolphins, he could ethically have replaced Kea and Puka with Phoenix and Ake, taken from the Gulf of Mexico. I said I could not imagine Sue Savage-Rumbaugh replacing a dead or kidnapped Kanzi the bonobo, Irene Pepperberg replacing Alex, Penny Patterson replacing Koko the gorilla, or Lyn Miles replacing Chantek the orangutan with a bonobo, African Grey parrot, gorilla, or orangutan captured for that purpose from the wild.[42] He never replied.

Phoenix and Ake were imprisoned in Gulfport for the first thirty days after their capture to acclimate them to living in a tank instead of the open sea, and to eating thawed, instead of fresh, fish.[43] Then they were shipped to Herman's lab. Over the next seven months, he socialized them to humans and prepared to begin their language comprehension training. They received forty-five grams of thawed smelt for every correct response they made to simple sound discrimination training tests, up to their daily ration of 8,500 grams.[44]

I said "language comprehension training." In the most infamous article in the annals of nonhuman language research, Herbert Terrace asked, in 1979, "Can an ape create a sentence?" He answered "No," and nearly destroyed the legitimacy of all nonhuman language research.[45] Herman asked a different question: "Can a dolphin *understand* a sentence?"[46] This paid dividends, for dolphins, like human children, and I, when I travel in Italy, can understand vastly more language than we can produce.[47]

Herman thought understanding human language requires one to know both its "syntactic" part (the information added when words are combined and shuffled according to grammatical rules) and its "semantic" part (the meaning and reference of words). These have been described as the "indispensable core of any human language."[48] The words and syntax Herman taught them, taken together, are their "languages."[49] Herman didn't teach the dolphins a known language and doesn't mean to imply that the English glosses he gives to inform us of what he's asked them to do signify they understand real English.[50]

Ake learned a "dolphinized" gestural language, distantly related to American Sign Language (ASL). Experienced teachers gestured to her at the rate of about a word a second and took a quarter second more to segue into the next word.[51] Phoenix was taught an acoustical language, a series of computer-generated whistles lasting a second or less.[52] Herman used whistles because dolphins naturally whistle. To forestall confusion in case a whistle already meant something to Phoenix, the whistles Herman chose didn't resemble natural ones. Whistle-words were produced about a quarter second apart.[53]

Before we dive into the dolphins' language work, let's look at their capacity for what Herman thinks is cognition's foundation: memory.[54] For a quarter century, dolphins have been known to have good working, or short-term, memory.[55] This is the ability to form, then maintain, a mental representation.[56] Kea could. After listening to eight tones, she was asked to indicate whether a ninth matched any of them. Just as humans do, she exhibited a classic "recency effect." She was better able to match the ninth tone to one of the eight the later in the sequence she heard it.[57] She matched hundreds of sounds to sample sounds after delays of up to two minutes.[58] Analogously, Phoenix matched objects she saw with sample objects. Herman concluded that dolphins have excellent short-term memory for things seen and sounds heard.[59]

Later, Herman taught Hiapo and the Elele, also captured in the Gulf of Mexico, the gestural word for "repeat." When the two saw him gesture "repeat," they were to repeat whatever behavior they had just completed. During training, they learned to repeat a small number of behaviors. But they also learned a generalized repeating rule that allowed them to repeat complex behaviors to which they were not trained.[60] Herman thought the simplest explanation was that they retained detailed internal representations of their recent actions in working memory.[61]

In June 2001, I received Elele's autopsy report. She had died on December 16, 2000, of clostridial enteritis and enterotoxemia.[62] The Dolphin Institute and KBMML issued a press release:

> Elele, whose name means "Ambassador" in Hawaiian, had been at KBMML since 1987. . . . The staff view their dolphin companions as ambassadors for their species, teaching us about their special nature, how they perceive their world, their intelligence, and how we may better communicate with them. Elele had contributed significantly to advances in the study and understanding of dolphin visual perception and vigilance, creativity, imitation, and self awareness.[63]

Herman asked that people "remember her intelligence, sensitivity, and energy, and her willingness to bridge the gap netween species through communication."[64]

Animal Rights Hawaii issued its own press release:

> The Kewalo dolphins are captives of the University of Hawai'i, which has sponsored Louis Herman's cruel research on the captured animals

> for 30 years. He . . . has subjected his prisoner animals to denial of simple pleasures—these animals confined to barren concrete tanks—too small, too shallow. Yes he wants to move . . . but he has been content with keeping his prisoners in inadequate sub-standard quarters for many years. . . . It is time for the surviving prisoners of TDI [The Dolphin Institute] to be returned, rehabilitated and released to their ancestral waters in the Gulf of Mexico.[65]

Herman is not alone in characterizing captive nonhuman animals as "ambassadors." The word is commonly used by zoos. One may argue, as Herman does, that his exploitation of dolphins is justified. But "exploitation" is an appropriate characterization of Herman's relationship to dolphins captured, then held captive at The Dolphin Institute. "Ambassador" does not correctly describe Phoenix or Ake. It did not describe Elele. Ambassadors represent their sovereigns. William Blackstone, English legal commentator of the eighteenth century, said that ambassadors "owe no subjection to any laws but those of their own country, their actions are not subject to the control of the private law of that state wherein they are appointed to reside. . . . An ambassador ought to be independent of every power except that by which he is sent."[66] "Captive" may seem a cruel word. But even Herman understood it to be accurate when he claimed to detect my "anti-captivity bias."

Dolphin Grammar

Can Phoenix and Ake understand two kinds of sentences? Herman instructed them to act with respect to some object and measured their comprehension by how accurately and reliably they did what they were told.[67] The dolphins could answer his "yes" or "no" questions by pressing a paddle on their left or right. For example, Ake could respond to BALL QUESTION (glossed as "Is there a ball in your tank?") by pressing the "yes" or "no" paddle. This tested their knowledge of what was and what was not in their physical world and helped answer the question: "Can symbols come to function as surrogates for their referents (or, more precisely, for the mental representations of those referents)?"[68]

Each dolphin learned about forty words, and not all the same ones. They included objects the dolphins could move (BALL), the trainers could move (NET), or those fixed to their tank (WINDOW). The dolphins might be required to act, OVER, UNDER, and TAIL-TOUCH; or told to use one object, called the "direct object," to do something to an "indirect object." There were two possibilities, FETCH (take the direct object to the indirect object) or place the direct object IN the indirect object. Herman taught them names for agents, PHOENIX and AKEAKAMAI; modifiers of objects, RIGHT, LEFT, SURFACE, BOTTOM; and several "control and function words," YES, NO, ERASE (cancel all previous words), and QUESTION (say whether an object is in the tank).[69]

At the core of their syntax is word order, as it is in English.[70] "Take the ball to the dog" does not mean the same thing as "Take the dog to the ball." Herman's rules are so complicated that I can't easily remember them and often have to refer to a chart.[71] He never explicitly taught grammar to either dolphin. Instead, they learned their languages implicitly, by working with grammatically correct word sentences, just as human children learn language. After working with two- and three-word sentences, the dolphins understood grammatically correct four-word sentences at once.[72]

Phoenix and Ake didn't always learn all the same rules of syntax. They did for two- or three-word sentences instructing them to interact with an object. Here are those rules: objects precede action and modifiers precede the objects modified. The sequence for a two-word sentence is object + action (WINDOW TAIL-TOUCH means "go to any of the underwater windows in your tank and touch it with your tail flukes"). Both understood this three-word sentence: modifier + object + action (LEFT BALL MOUTH, or "go to the ball on your left and place your mouth around it").

At three-word sentences, the grammars diverged. Phoenix learned a linear, left-to-right grammar, similar to that of English, and she has to act in the order of the words given, left to right.[73] She might be told to take one object to another (HOOP FETCH PIPE), or place one object into another (PIPE IN BASKET). Herman taught this four-word combination: modifier + direct object + action + indirect object (BOTTOM HOOP FETCH BASKET, or "go to the hoop lying on the bottom of your tank and take it to the basket").

Herman invented a second four-word rule for Phoenix: direct object + action + modifier + indirect object (PIPE FETCH BOTTOM BASKET, or "take the pipe to the basket lying at the bottom of your tank"), by combining two rules for the three-word sentences she already knew: modifier + object + action and direct object + action + indirect object. Phoenix correctly responded to this new four-word sentence on her first try and to nearly two-thirds of them presented over several months.

Ake, on the other hand, learned a nonlinear "inverse grammar." Her action sequences were performed in the order opposite to how she received them.[74] One three-word rule of syntax was indirect object + direct object + action. PIPE HOOP FETCH told her to "go to the hoop and take it to the pipe." This meant she had to wait for the entire instruction before acting. As he did with Phoenix, Herman made up a four-word rule by combining two of the three-word rules he had taught her: modifier + object + action and indirect object + direct object + action. The resulting rule was modifier + indirect object + direct object + action (RIGHT BASKET PIPE FETCH, or "take the pipe to the basket on your right"). Ake got this sentence correct on her first attempt and more than three-quarters of them right over the next four months. She also responded perfectly to a different four-word rule, indirect object + modifier + direct object + action (SPEAKER LEFT HOOP FETCH, or "go to the hoop and take it to the left speaker"), which combined the same two familiar rules in a different way, the first time it was presented and on more than half the times she saw them over the next four months.

Both learned five-word sentences, too. Phoenix's followed this rule: modifier + direct object + action + modifier + indirect object (BOT-

TOM BASKET FETCH BOTTOM HOOP, or "take the basket on the bottom of the tank to the hoop on the bottom of the tank"). Ake's obeyed a different rule: modifier + indirect object + action + modifier + direct object (RIGHT HOOP LEFT PIPE FETCH, or "take the pipe on the left to the hoop on the right").[75]

Herman is especially impressed with Ake's and Phoenix's ability to understand sentences whose meanings reverse when word positions reverse. Ake can tell the difference between BASKET BALL FETCH ("go to the ball and take it to the basket") and BALL BASKET FETCH ("go to the basket and take it to the ball"). In one test, Ake responded correctly to 59 percent of three- and four-word "reversible sentences," far above chance. Phoenix did even better. She got 77 percent of three-, four-, and five-word reversible sentences right.[76] One way to demonstrate that an animal understands sentences is to see whether she can respond correctly to grammatically correct novel sentences.[77] More than half the reversible sentences for both dolphins were ones they had never experienced.

Significantly, when Ake blundered, her errors never implied she didn't understand syntax. She might take a pipe to the basket when asked to take a pipe to the ball. But she never took a ball to the pipe. Or she might take the right pipe to the ball when asked to take the left pipe. Phoenix made just one of these mistakes.[78] Herman concluded the "extreme rarity of reversal errors confirms the sensitivity of the dolphins to the word-order rules of their respective languages."[79] Keep in mind that an incorrect answer doesn't necessarily mean a dolphin doesn't know the correct one. Karen Pryor reports that one dolphin, for whom correct answers were reinforced with fish dispensed from a machine, responded correctly to an experiment over many days. To her surprise, the dolphin suddenly began to get everything wrong. It turned out the fishes in the machine had dried out and become inedible, at least to a dolphin. When they were replaced, so were the wrong answers.[80]

Different placements of the same words in a sentence might result in new meanings other than a simple reversal. Told WATER RIGHT BALL FETCH, Ake was to go to the ball on her right and take it to the thin jet of water streaming from a hose. Told RIGHT WATER BALL FETCH, she was to go to any ball and take it to the stream of water flowing to her right. Phoenix might be instructed BOTTOM HOOP FETCH SURFACE FRISBEE. That meant "go to the Frisbee at the bottom of your tank and take it to the hoop at the surface." Or she might be told SURFACE HOOP FETCH BOTTOM FRISBEE, "go to the hoop at the surface and take it to the Frisbee at the bottom."[81] Given four-word sentences of indirect object + modifier + direct object + action, such as BALL RIGHT FRISBEE FETCH, Ake understood that it was the right ball and not the right Frisbee to which she was supposed to attend.[82]

Understanding sentences underscored what Herman called the dolphins' "sense of grammaticality."[83] As they learned new words and showed they understood simple rules, Herman gave the dolphins increasingly difficult sentences to understand.[84] By 1987, these had accumulated to more than a thousand sentences; by 1990, more than 2,000.[85]

Making Sense of Nonsense

Ake and Phoenix demonstrated that they were aware of which syntax and word meanings they knew and which they didn't. Herman sometimes presented them with either grammatically correct sentences, similar to Noam Chomsky's famous nonsense "Colorless green ideas sleep furiously," that contained one, sometimes two, words they didn't know, or with grammatically senseless sentences. Told PIPE PHOENIX FETCH ("take Phoenix to a pipe"), Ake began to swim on seeing PIPE, to return at PHOENIX, who had no intention of being dragged to a pipe, or anywhere else. Phoenix might be asked, in a grammatically correct way, to move a fixed object: WINDOW FETCH HOOP ("take one of the underwater windows in your tank to the hoop"). She might swap an object she could carry for WINDOW and bring that to the hoop.[86]

Occasionally Ake repaired a nonsensical sentence by responding properly to meaningful segments within it. Told PIPE NET SPIT ("spit the net on the pipe"), she ignored PIPE and spit on the net.[87] Instructed PHOENIX HOOP FRISBEE FETCH, she would usually carry out one of the three subsets of the sentence that made sense, either PHOENIX HOOP FETCH ("go to the hoop and take it to Phoenix"), PHOENIX FRISBEE FETCH ("go to the Frisbee and take it to Phoenix"), or HOOP FRISBEE FETCH ("go to the Frisbee and take it to the hoop"). This demanded she analyze the nonsensical sentence backwards and forward to pick out the usable fragments. She never repaired a sentence by assuming the grammar was incorrect. She might reject a sentence such as BASKET WINDOW FETCH ("take one of the windows in your tank to the basket"). But she never brought a basket to the speaker.[88]

Rejection and repair were, Herman says, by themselves "conceptual leaps."[89] Sometimes what seemed nonsense to Herman made sense to Ake. Testing Ake's responses to nonsense sentences, Herman instructed her to WATER TOSS. WATER referred to water that streamed from a hose into her tank. TOSS instructed Ake to throw the named object into the air. WATER TOSS was therefore supposed to be senseless, for Ake couldn't toss a stream of water into the air. Or could she? Ake swam to the hose, repeatedly jerked her head through the water, and threw out a large spray. Presented with the same sentence, Phoenix responded the same way.[90]

Phoenix and Ake could show only that they knew an answer or they didn't. What if they were uncertain? Another team of scientists investigated this. They gave tests to a single dolphin and a pod of college undergraduates. It asked them to categorize a tone as either "higher" or "lower" than a sample tone by pressing one of two paddles (the "higher" or "lower" paddle). If uncertain, either dolphin or human could avoid answering by pressing a third paddle. The higher tones were slowly lowered, the lower tones slowly raised. As the tones began to fade one into another, dolphin and undergraduates alike began to hit the escape paddle. The dolphin would swim slowly toward the third paddle and wave her head before she pressed. Overall, the circumstances under which either dolphin or undergraduate escaped were similar. Under questioning, the undergraduates explained their behavior as uncertainty ("I wasn't sure").[91]

Every natural language permits "recursion." It allows sentences to grow larger and more complex. Phoenix was tested on a kind of recursion for which she was never trained, the conjoining of sentences. She was told to act on one object twice: object + action + object + action. An example was PIPE OVER PIPE TOSS, meaning "jump over the pipe, then toss the pipe." She usually succeeded.[92]

When told BALL HOOP OVER ("jump over the ball hoop"), Ake would jump over the ball, jump over the hoop, or reject the nonsense altogether.[93] Told BALL ERASE HOOP OVER ("ignore the ball and jump over the hoop"), she would do it. Later, she was taught another recursive rule, the conjunction AND. BALL AND HOOP OVER demanded she jump over both a ball and a hoop. SPIT AND WATER OVER required her to spit water while she was leaping.[94] And that was what she almost always did.[95]

How the Dolphins' Words Symbolized the World

How do we know the symbols in the languages taught to Phoenix and Ake represent real objects or real events happening in the world outside their minds? Herman argues from three lines of evidence. They are the dolphins' ability (1) to generalize symbols to situations different from the one in which they learned their symbols, (2) to find objects not visible or present when told to do something with them, and (3) to report objects that are not in their tank.[96]

The dolphins were trained in ways that helped them learn to generalize words and sentences beyond the specific objects and places used to teach them to label.[97] For example, they learned HOOP with a large eight-sided hoop made of plastic pipe. Later, they used square hoops, round hoops, big hoops, little hoops, dark hoops, white hoops, thick hoops, thin hoops, floating hoops, and sinking hoops, all of which the dolphins recognized as a HOOP. Similarly, when her tank needed repair, Ake was shifted to a temporary tank. She had no trouble identifying the waterfall cascading into her temporary tank as the same WATER that streamed from the narrow hose in her regular tank.[98] The dolphins generalized PIPE, BASKET, BALL, and FRISBEE the same way.[99]

Ake and Phoenix did the same thing with action words. Phoenix was taught THRU using a hoop (HOOP THRU). When presented the novel sentence GATE THRU, she immediately swam through the open gate in her tank. Instructed GATE THRU when the gate was closed, she swam to the gate, paused, pushed it open, then entered. The first time Ake was told to place a ball IN a basket, a ball was already there. She pulled it out, swam away with it, then returned and plopped it back into the net.[100]

Ake might be told UNDER FETCH, when no ball was in her tank. After a delay of thirty seconds, Herman would throw in a ball. Ake usually swam under it.[101] During the delay, she might even rehearse by beginning to swim on her back, the position in which she swam beneath objects. Told PECTORAL TOUCH the object, during the delay she might thrust her pectoral fin out from her body, which is how she touched things with the fin.[102] She also understood LEFT and RIGHT means her left and right, wherever she is, as Phoenix understood SURFACE and BOTTOM are specific locations no matter where she is.[103]

To allow Ake to report on whether an object was in her tank, Herman taught her to press a NO paddle when he pointed to it. Then he gave her a two-word sentence—say, BALL FETCH—when no ball was in her tank. She searched. She gave up. Then Herman pointed to the NO paddle. She pressed it and received fish as a reward. That was all the training she needed to use the NO paddle successfully from then on.

She would search until, satisfied the object was not in the tank, she would press the NO paddle more than 80 percent of the time; the other 20 percent, she would do something to an object in her tank.[104] Later, Herman taught her to press a YES paddle. He would throw up to seven objects into her tank and ask whether a certain object, say, a ball, was among them (BALL QUESTION). She would press the YES paddle if it was. Or he would tell her to do something to an object (BALL UNDER). If the ball was present, she would swim under it. If the ball was not there, she hit the NO paddle. She was right more than 80 percent of the time.[105] Asked to take a ball to a nonexistent basket, Ake took it to the NO paddle. When both ball and basket were in her tank, she either placed the ball into the basket or brought it to the YES paddle. If the ball wasn't there, she pressed the NO paddle.[106]

Both dolphins could make nearly impossible instructions possible to perform. When an object over which they were to leap drifted or blew to one side of the tank, they would push the object away from the wall, then leap over. Told BOTTOM HOOP THROUGH for the first time, when the only hoop in the tank was lying flat on the bottom, Phoenix lifted the hoop until it was vertical, then swam through. Ake did the same thing when bid HOOP THROUGH. Told SURFBOARD FRISBEE IN, she placed a Frisbee atop the surfboard. It rolled off. She replaced it. It fell again. She replaced it again and kept replacing it.[107]

Finally, Phoenix responded entirely correctly to 138 of 207 novel sentences, Ake to 220 of 372, sprinkled along with dozens of sentences to which the dolphins had been exposed at least once.[108] Even their mistakes showed they almost always understood the thrust of what they were being asked to do, for they rarely, just 13 percent of the time, made more than one mistake.[109] When Ake was told NET LEFT BALL FETCH ("take the ball to your left over to the net"), she erred by bringing the ball on her left to a human instead of the net.[110] Phoenix also easily substituted an object she was shown, say, a ball, for its symbol in a sentence of her acoustical language. Herman might sound the acoustical symbol for FETCH and she would fetch a ball.[111]

From this and such evidence as Ake being taught IN (to place one object into another), then using IN properly the first time Herman used it in a sentence, Herman concluded the dolphins "knew the semantic boundaries of their world."[112] He sees the following as strong evidence the dolphins learned "an internal representation of a symbol sequence": Ake's ability to understand four-word sequences immediately; Phoenix's comprehension of conjoined sentences; Ake's knowing how to use IN within a sentence on her first try; her ability to retrieve a grammatical string of words from within longer ungrammatical sentences; and her responding accurately to a sentence when its sequence of words doesn't parallel her sequence of actions.[113] Evidence powerfully suggests dolphins understand that symbols are referential and that the answer to the question "Can a dolphin *understand* a sentence?" is "Yes."[114] That Phoenix uses her acoustical language about as well as Ake uses her gestural language implies that dolphins have a general ability to understand language, not limited to what they see or hear.[115]

Spontaneously Understanding Signs

We know that after hard work, dolphins can learn to understand symbols. But with dogs and chimpanzees, can they spontaneously comprehend such communicative signs as pointing and gaze following? Three experiments plumbed Phoenix's understanding of pointing. First, Herman placed three familiar objects so they formed the points of an imaginary equilateral triangle twenty-two feet from Phoenix and a scientist sitting on a surfboard beside her in the center of the tank. The surfboarder would point for a second or two with his hand and arm to one of the objects, then sign OVER. Eighty-one percent of the time, Phoenix swam to the object to which he pointed and leaped over.[116] When this test was repeated, she got 89 percent right. In a second experiment, a surfboard, a basket, and a hoop were arrayed to Phoenix's right, rear, and left. A scientist facing Phoenix pointed to one object for a second or two with her arm and index finger, signed FETCH, then pointed to a second object, intending to tell her to "take the first object to the second." Phoenix complied half the time.[117]

A third experiment was performed with both Ake and Phoenix. Herman chose three from the following four objects, a basket, a surfboard, a pipe, and a hoop, and placed them to Ake's right, left, and rear. The ones to her side were about twenty-five feet away; the one to her rear, more than forty feet. Herman then pointed to one of the three objects with his arm and index finger for less than a second and followed the point with an action sign such as FETCH; or he didn't point at all, but used signs for both object and action, HOOP FETCH. The only trouble Ake had was in responding to Herman's pointing to the object behind her. Phoenix had no trouble understanding Herman's pointing to objects on either side, either. But she also had trouble, though far less than Ake, understanding his pointing to objects to her rear.

Herman tried moving all three objects closer, to within eight to ten feet, to see how Ake would react. She still had problems with the object behind her. Herman thought he might be able to teach her to understand his pointing to her rear if he greatly exaggerated his arm movements when he pointed and leaned into the direction of his pointing. She got it. And continued to get it, even as Herman gradually lessened the exaggeration until his pointing was back to where it had begun.[118]

Now things got complicated. Remember, Ake's was an inverse grammar; her action sequences were performed in the order opposite from which they were given. Would Ake take a pipe to a hoop when Herman pointed to a hoop, then flashed PIPE FETCH? She did, 89 percent of the time. What if Herman ordered HOOP, pointed to a pipe, then signed FETCH? Now she took the pipe to the hoop 61 percent of the time. What if Herman pointed first to the hoop, then to the pipe, and signed FETCH? This time she complied 72 percent of the time.[119] There was no doubt that Ake understood pointing as a reference to objects.[120]

Dolphin Imitation

Without doubt, dolphins imitate and can do it throughout their lives. They imitate the actions of other dolphins. One bottle-nosed dolphin imitated the spinning leap of a Pacific spinner dolphin at the action level immediately after seeing it for the first time.[121] Herman gives an example of Ake's and Phoenix's program-level imitation. After watching humans throw Frisbees, the dolphins started tossing them from their mouths.[122]

They imitate vocalizations.[123] In the early 1990s, dolphins learned to associate computer-generated whistles from an underwater keyboard with objects and actions and to produce short novel word combinations when playing with two objects.[124] Herman trained dolphins to mimic electronically generated whistles.[125] Amazingly, one captive dolphin working with Diana Reiss imitated a complex half-second whistle by copying the last part of the whistle first, then the middle part, then the beginning, all in the following half second! Several dolphins learned to associate whistles with objects such as balls or floats. They could press the keys of an underwater keyboard to produce one whistle for a ball, another for a float, others for one of the other objects. When they did, Reiss threw that object into the tank. The dolphins began spontaneously to imitate these computer whistles and use them appropriately, for example, often whistling "ball" while playing with balls.[126]

Perhaps it shouldn't surprise us that Peter Tyack, a marine mammal biologist, finds captive dolphins "fantastic imitators of human sounds" as well, able to imitate sounds they have never heard before, and within seconds.[127] He thinks that every wild dolphin develops a unique "signature whistle." Dolphins live in complex, swiftly changing societies. Because they whistle primarily when out of each other's sight, scientists think that by using whistles, dolphins "are communicating potentially complex information" to recognize each other and keep their groups together.[128] Wild dolphins also imitate the whistles of other dolphins, apparently as a way of addressing them as friend or foe, which may make them linguistic symbols.[129]

Herman and Ake did a little work with imitation, too. As a prelude to teaching Ake to produce words, Herman trained her by a method used to teach imitative speech to mute schizophrenic children. Then he tested to see whether she could mimic computer-produced sounds broadcast through an underwater speaker. When she succeeded, she heard the sound for YES. She heard the sound for AKE if she failed.[130] She soon began to mimic every sound broadcast to her, often mimicking novel sounds the first time she heard them, and producing sounds so close to the original that Herman could hardly distinguish her sound from the computer's, even when using a recorder.[131] Within a few months, she could use her whistles to label HOOP, PERSON, BALL, and three other objects.[132]

Dolphin Self-Awareness

Dolphins have no arms or hands or fingers. They live in a featureless, trackless aquatic world scarcely resembling the land we inhabit and their ancestors abandoned. Their hearing is far superior to ours. They have no sense of smell. But their echolocation is so sophisticated they can visually choose an object they previously only echolocated.[133] It's an ongoing challenge to adapt to dolphins what we know about self-awareness in visual land-dwelling primates, who have hands and legs and who can't echolocate.[134] Ken Marten and Suchi Psarakos, leaders in investigating marine mammal self-awareness, are uneasy about assuming that the "sense of self will be the same in dolphins as in primates or procedures drawn from the study of primates will reveal it in dolphins."[135] Perhaps dolphins incline toward a self-recognition that depends on what they hear or echolocate rather than what they see.[136]

Scientists have used Gordon Gallup's standard mirror self-recognition (MSR) test with dolphins. One team included Gallup himself. Finding the results hard to interpret, team members thought they might have focused too narrowly on technically replicating the test rather than adapting it to dolphins.[137] Marten and Psarakos used zinc oxide to mark the sides of Keola, a fourteen-year-old male, Itsi Bitsi, a thirteen-year-old female, and three Pacific bottle-nosed dolphins *(Tursiops gilli).* The dolphins were unlikely to see the zinc oxide marks on their sides without a mirror, a large piece of reflective mylar the researchers stuck to a tank window. It also functioned as a one-way mirror through which Marten and Psarakos filmed the dolphins' reactions. They thought the dolphins used it to examine their marks if they turned or twisted to bring it into view within ten minutes of marking and looked at the mark for at least ten seconds from a distance of one foot or less. Those most intimate with the dolphins had to agree that the dolphins were using the mirror to examine the marks; and all did agree that at least Keola used the mirror.[138]

Marten and Psarakos pointed a video camera into the tank near a television monitor allowing the dolphins to watch themselves in real time; then they played the videotape for the dolphins to watch. They hypothesized that self-aware dolphins would behave differently when looking at themselves in real time than on playback. That was just what Keola did. He spent much time opening his mouth wide and rhythmically moving his head when looking at himself in real time, but not on playback.[139] Marten and Psarakos concluded that although they hadn't proved Keola's MSR, his case was strong.[140]

The case has since become compelling. In 2000, Diana Reiss and Lori Marino tested two Atlantic bottle-nosed dolphins at the New York Aquarium. Marked with temporary black ink on body parts they could see only with a mirror, the dolphins were videotaped as they interacted with a Plexiglas mirror and the glass walls of their tank. Two independent observers who rated the dolphins' behavior had to agree on what they saw for any behavior to count as mirror self-recognition. After analyzing the results, Reiss and Marino found "definitive evidence" of MSR.[141]

We'll see in Chapters 10 and 11 that like chimpanzees, orangutans and gorillas passed Gallup's MSR test. Because dolphin and primate brains are wired so differently, Marino and Reiss believe that dolphin MSR "may be based on a different neurological substrate."[142] They were intrigued by the similarity of the dolphin and dog results. They agreed that dogs might have been selectively bred to pay close attention to human signals. Not dolphins. They speculated that dolphins redirect the brain power used to communicate with other dolphins to read and predict human communication and that "various species may be capable of solving the same sorts of problems using different cognitive mechanisms."[143]

Theories of Dolphin Intelligence

Some theories of the evolution of intelligence stress tool use as the engine.[144] Except for some Australian dolphins who carry sponges, apparently to protect their rostrums (or beaks) as they forage on the sea floor, there seems to be little dolphin tool use.[145] Dolphins, like chimpanzees, present strong evidence that living in social groups spurs intelligence.[146] Like chimpanzees, dolphins live in complex fission-fusion societies in which individuals associate in groups having a changing membership. But the comparative rates of change between dolphin and chimpanzee groups resemble the comparative rates of action between the Keystone Kops and the diners in "My Dinner with Andre." Dolphin groups may change hourly, sometimes several times an hour.[147]

Male dolphins form exceedingly complex long-term coalitions of varying strengths and sizes that sometimes temporarily combine to form alliances for purposes of aggression and defense.[148] Large groups can cooperate. Sometimes when a small group of dolphins encounters a large school of anchovies they signal their find to other dolphins and wait for help. Sometimes reinforced by hundreds of others, the dolphins herd the fish to the surface and hold them there; they then take turns swimming through the school, eating. If the school splits into smaller units, some of the dolphins will chase them back into the main body.[149]

Dolphins sometimes cooperate with humans. For perhaps a century, a small group of dolphins off the coast of Laguna, Brazil, have daily signaled the presence of fish to men standing in water so cloudy that they cannot see the fish. The dolphins' rolling dives, fifteen to twenty feet from the men, signal not just the presence of fish but the direction in which they are swimming. At the dolphins' signal, the fishermen cast heavy nets that stun and trap the unseen fish. The dolphins dine on the disoriented fish the nets miss.[150] Off West Africa and Australia, fishermen cast their nets and then beat the water. Hearing the sound, dolphins herd schools of mullet toward the nets.[151]

Males with sturdy bonds may swim close together side by side, in synchrony, as may mothers and young. They rub against and stroke each other with their pectoral fins or swim with one dolphin resting his pectoral fin against the other.[152] Calves spend years with their mothers.[153] Says biologist Rachel Smolker, "Like us, dolphins spend much of their brainpower keeping track of who does what with whom, engaging in rivalries and social politics, figuring out what others might be thinking, competing and cooperating in complicated, multi-leveled alliance. Like us, their minds are on each other." Having "watched the intensity and complexity of dolphins' social interactions" in Australia for almost a decade, Smolker is "convinced that this is the driving force behind dolphin intelligence."[154]

Rocky

On a bright warm autumn day, I met Ron Schusterman at his office in the Long Marine Laboratory at the University of California, Santa Cruz. His building teeters on the edge of a cliff that soars from the north shore of Monterey Bay, midway into a spectacular concave sweep of mile after mile of ocean and rocky headlands. He began his career by investigating the cognition of chimpanzees, gibbons, and monkeys, then turned his interest to dolphins and pinnipeds (the suborder that includes seals and walruses). He took a few years to study the echolocation of dolphins in Hawaii and observed Louis Herman's work. Now he directs the university's Pinniped Research in Cognition and Sensory Systems program.

Over bean burritos at a Mexican cafe in Santa Cruz, Schusterman told me that he found dolphins "smarter" than pinnipeds. As Rachel Smolker does, he suspects that their complex and intense social relationships probably account for their intelligence. Sea lions, he said, are more like "wet dogs." I asked about his criticism of Louis Herman's work: he thought Herman was "more liberal" in his data interpretations than he was.[155] Still, when I met Schusterman's prize student, Rocky, a California sea lion *(Zalophus californianus)* that afternoon, I reflected that she had done some amazing work, as had other sea lions.

Rocky is a female, twenty-four years old. She has lived at the Long Marine Laboratory for almost her entire life, spending much of her time in a concrete tank 7.7 meters (about 25 feet) in diameter and 1.8 meters (about six feet) deep.[156] Schusterman taught Rocky a gestural language similar to Ake's. Within two and a half years, Rocky learned signs for eleven objects, such as balls, pipes, and cubes; five modifiers, such as size and brightness; and six actions signs, including FETCH. Using an "inverse grammar" similar to Ake's, Rocky could comprehend any one of more than 7,000 sentences of up to seven signs, such as SMALL BLACK CUBE LARGE WHITE BALL FETCH ("go to the large white ball and take it to the small black cube").[157] Rocky's ability to apply the relations between words and even to unfamiliar sentences showed that she had acquired a syntax.[158]

At Louis Herman's laboratory, Huapala, a young sea lion, learned the concept of identity (understanding that two things are the same), then transferred it to new objects. She could retain a short-time memory to match objects for at least a minute about 70 percent of the time, an ability just below that of dolphins and monkeys.[159] Schusterman sees categorization as a powerful tool that brains invented to simplify a complex world; it requires an ability to lump things that may not look alike into a category when they function the same way.[160] He trained Rocky and a second sea lion, Rio, to understand the two classes, "letters" and "numbers." Rocky and Rio then learned to match letters to letters (rejecting numbers) and numbers to numbers (rejecting letters). Eventually, they could match letters they had never seen with letters they already knew, and match novel to known numbers; in other words, letters to letters and numbers to numbers. Schusterman thinks this allows sea lions, and other nonhuman animals, to recognize not just individuals but relationships between individuals and relationships between individuals and groups, all of which should help give us "an appreciation for the sophisticated ways animals think without words."[161]

Do Phoenix and Ake Have Practical Autonomy?

With Phoenix and Ake, we encounter our first Category One animals outside of chimpanzees and bonobos. They so clearly possess practical autonomy that I assign them an autonomy value of 0.90, higher than that of Alex the parrot, primarily because they possess selves, both James's "Me-self" and "I-self," probably Damasio's autobiographical self, and not just Neisser's ecological and interpersonal selves, but at least a conceptual self. Atlantic bottle-nosed dolphins clearly demonstrated MSR in at least one carefully run experiment and have acted in a manner consistent with MSR in several others. They undoubtedly understand complicated human sentences and possess a grammar. They know what language and syntax they understand and know what they don't; they can also separate sense from nonsense. Any creature this advanced in language-like communication routinely acts intentionally and forms sophisticated mental images. Dolphins have concepts and spontaneously understand human pointing, gazing, and the holding up of replicas. They instantly imitate actions and vocalizations.

Few physical, economic, political, and psychological obstacles block their attainment of basic dignity-rights. They have little economic value. They are not eaten, generally used in biomedical research, or any human endeavor other than entertainment, and they are few in numbers. They are loved by the public, and are believed to be very mentally sophisticated. In one study, between 90 percent and 98 percent believed dolphins experience emotions and pain, are intelligent, reason, suffer, and are self-aware. In intelligence, consciousness, and ability to experience pain and suffer, the public ranked them at about the level of chimpanzees.[162]

NINE

Echo

Amboseli

Cynthia Moss's books and BBC documentaries have made Echo the most famous living elephant in the world.[1] She is the matriarch of the EB family in Amboseli National Park, which lies in the enormous shadow of Mt. Kilimanjaro, almost four miles high and the signature of that part of East Africa. I went to Kenya to meet Echo. She is an African elephant, a mammal of the order *proboscida,* family *elephantidae,* genus *Loxidonta,* species *africana.* Male African elephants can stand thirteen feet at the shoulder and weigh seven tons; one gigantic bull was said to have a shoulder height of thirteen feet and a weight of eleven tons.[2] Echo stands about eight feet at the shoulder and weighs seven or eight thousand pounds.[3]

Her gigantic brain weighs ten to twelve pounds; at birth, it was 35 percent its present size. Its volume is 4,500 to 5,000 cubic centimeters. She has a formidable frontal lobe, and her cerebrum's temporal lobes, a primary storage area for human memory, are so large that they bulge, and are even more convoluted than mine.[4] As always, I asked Mr. Darwin when Echo and I last shared an ancestor. Same as for dogs, he answered, about 90 million years ago.[5]

When she arrived at Amboseli in 1972, Cynthia Moss had elephant-watched with Iain Douglas-Hamilton, the Jane Goodall of elephants, at Lake Manyara National Park in Tanzania on and off for four years. Amboseli now holds 1,100 elephants, and Moss knows them all.[6] As most of Amboseli is like a pancake and, at 150 square miles, one of the smallest national parks in Africa, I figured I had a fair chance of finding Echo, even though Moss's books and the films occasionally mentioned how hard that could be.

Our driver, Stanley Kinyolo, has been navigating Amboseli for nine years and had an idea where Moss camped. On May 31, 2001, together with two Vermont filmmakers, Paul Garstki and Donna Thomas, I jolted past wildebeests, zebras, buffaloes, warthogs, and jackals, all of whom paid us little heed, our ancient Range Rover adding dirty clouds to the dust devils swirling the land. The park gets so little rain it's technically a semiarid area, nearly desert; but oddly, and luckily for its animals, it's a swampy desert, for the snows of Kilimanjaro are always melting. At 8:00 A.M., Stanley turned left at a den of hyenas, ignored a "No Entry" sign, and we were there.

Moss was breakfasting with an old friend, Sandy Price, and two German guests inside a large dining tent. It was lined by ninety-two sun-bleached lower elephant jaws and one large skull belonging to Stephen, a bull elephant. Moss keeps all the jaws she finds, for they help her age the elephants at death. Among them, she said, were the remains of many an old friend. That morning, she was much worried about Alfred, a large bull whom she had known almost from the beginning of her life at Amboseli. He had been foaming at the mouth, and she had no idea whether he had contracted a strange, perhaps contagious disease. That would be unusual, she said, for aside from suffering heart disease, elephants rarely fall ill. Veterinarians from the University of Nairobi were on their way. An hour later, we piled into Moss's Land Cruiser and began our search for Echo. We were hopeful; Soila Sayialel, a local Maasai woman who is Moss's project manager and who has worked closely with Moss for fifteen years, had spotted Echo just the day before in the northern portion of the nearby Longinye Swamp.

Iain Douglas-Hamilton discovered that Manyara's adult female elephants, or cows, and their youngsters lived in families run by the eldest female. The older males lived apart. Moss found Amboseli's elephants living the same way. Sometimes more than thirty, sometimes just two, the families averaged seven when she arrived in Amboseli, fifteen today.[7] Moss double-lettered each family. The first, photographed in 1972, she dubbed the AA family and gave each a name beginning with the letter "A": Annabel, Amy, Alison.[8] Fifty-three elephant families live in Amboseli today.[9]

Related families who spend most of their time together form what Moss called a bond group and Iain Douglas-Hamilton called a kinship group.[10] The EA family, often found with Echo's EBs, form a bond group.[11] Moss thinks bond groups are usually extended families that split when they grew too large, though she knows at least one made up of unrelated families.[12] In 1974, she concluded that four small families, totaling twenty-three elephants, formed a single bond group, and named them the TA, TB, TC, and TD families.[13] Her 1988 book, *Elephant Memories,* is the story of their lives.

Moss gave the name "clan" to bond groups who use the same area during dry season.[14] With the four T families and the DA, DB, and SA families, Echo's bond group originally formed the Ol Tukai Orok clan, for they spent the dry season around the Longinye swamp and the Ol Tukai Orok woodlands by day and areas south and east by night.[15] Over the years, they drifted away, as the TC and TD families moved to the Kimana Sanctuary and the SA family to the west of Amboseli.[16] Acoustic biologist Katy Payne, who has long worked at Amboseli, discovered that elephants often communicate by sound below the level of human hearing. She has summed up the essence of the family, the bond group, and the clan: families are genetic, bond groups behavioral, clans geographical.[17]

Elephant families are large emotional knots. Females may die without ever having been alone.[18] They move, eat, and sleep together, usually in close proximity, often touching. They care for each other's children and present a common defense.[19] At danger, the family will bunch into a defensive circle or semicircle, adults facing out.[20] Katy Payne thinks if elephants are self-conscious, females have a sense of themselves as community members.[21] They "live empathetically: the experiences of others become their own experiences." But the males' sense is probably of themselves as individuals.[22] Moss suspects males, compelled to live a more isolated existence, experience loneliness.[23]

Echo, who was born in the final year of World War II, Moss estimates, has for decades been the undisputed matriarch of the EBs. She may live another ten years.[24] Since Moss first encountered the EBs in August 1973, they have always been aware of where Echo was "and what she was doing. If they were resting and she woke up and moved off, they would move off. If there was a smell or sound of danger, they would look at her first and then act. If she called them with a low rumbling vocalization, they would come, and if she made the 'let's go' rumble they would follow. She was their core, their anchor, their leader."[25] The matriarch is at the center of every elephant family. Built on three generations, the EB family has twenty-seven members. Echo has five daughters, Erin, Enid, Eliot, Ebony, and Emily Kate, and a son, Ely. Erin has given Echo four granddaughters and two grandsons not yet independent bulls.[26] Eleanor has a son, Enid a daughter, and both Erin and Edwina have a son and daughter. Thirty-six-year-old Ella is either Echo's daughter or sister and has given birth to two males and two females. Emily, who also may have been Echo's sister, was alive when Moss first encountered the EBs but died in 1989, leaving Eudora; she gave birth to a daughter, who now has her own boy, and a son.

Five minutes into our search, we encountered the PC family. Over thirty elephants, they grazed a hundred feet from the road, two subfamilies separated by fifty feet. One half was led by Patricia, the other by the single-tusked matriarch, Phoebe. Moss eased into the gap. They hardly noticed. We didn't move for an hour and a half, quietly chatting, watching, and snapping photographs; prancing curious babies, big sisters, preadolescent males, and massive females slowly and without concern ate their way around us.

Almost as soon as we left the PCs, we stumbled upon sixty more elephants crossing our narrow gravel road in a line that extended half a mile. These were the BB and UA families, part of the same bond group, and the PAs, of the same clan. No Echo. After photographing and observing, we returned to camp. There, grazing just to one side of the dining tent, was the famous TA family. Numbering six in 1974, they grew to eight in 1986, and eleven in the summer of 2001. I could see seven: Tillie (after I left, Tillie was killed by a poacher), Tillie's daughters, Tefilah and Tulip, her sister, Tecla, her daughter, Teryl, a male calf, and the matriarch Tonie's son, Tobyll. Tonie, her two young sons, and Tillie's one-year-old daughter (who will probably die without her mother) were almost certainly grazing in the nearby trees.

We set out again that afternoon. As we were leaving, Moss received the news by radio of Alfred's death. It visibly affected her. He had not been diseased, but tusked through the mouth by another bull. The festering wound had formed a huge abscess that sickened him, and he was starving. Mercifully, he had been anesthetized, then shot.

A sad Moss took us through miles of swamps and over desert, scouring Echo's afternoon haunts. No luck. She apologized again and again. But she reminded me that when she tried to find Echo for filming, she often searched fruitlessly for hours, even days. Sometimes when she returned to camp in frustration, there Echo would be. Apart from the TAs, only the EBs tended to come to camp.

"Very Echo-like," Moss said.

My hopes rose when, near camp, we encountered the EA family, the other half of Echo's bond group, grazing in the nearby swamp. As we turned at the hyena den, I strained to see Echo.

Elephant Emotions

Everyone who studies elephants marvels at how often and strenuously family and bond group members greet, even after short separations. Moss and behavioral ecologist Joyce Poole, who has studied Amboseli's elephants nearly as long as Moss has, report hundreds of these reunions. Families may rush together in great excitement, rumbling, trumpeting, shoving, bumping, swaying, backing into one another, trunk-twining, ear-flapping, spinning, sniffing, defecating, urinating, clicking tusks, reaching their trunks into the mouths of others, and secreting profusely from unique temporal glands on the sides of their heads. This pandemonium may last for ten minutes. The more intense the greeting, the closer the relationship among the greeters.[27] However, greetings need not be reserved for family. Zookeepers may be greeted, and Poole believes she has twice been greeted after long absences.[28]

Charles Darwin credited reports that told of Indian elephants weeping in Ceylon after being captured and bound, and of elephants weeping at the London Zoological Gardens when deprived of their children.[29] Scientists, elephant trainers, zookeepers, hunters, and novelists report that elephants cry when punished, shot, injured, or giving birth.[30] The unanimity and depth of certainty of the world's most experienced elephant-watchers that their subjects are emotional is remarkable. Iain Douglas-Hamilton has "little doubt that when one of their number dies and the bonds of a lifetime are severed, elephants have a feeling similar to the one we call grief."[31] Joyce Poole is "confident that elephants feel some emotions that we do not, and vice-versa [and that] we experience many emotions in common."[32] She writes:

> It is hard to watch elephants' remarkable behavior during a family or bond group greeting ceremony, the birth of a new family member, a playful interaction, the mating of a relative, the rescue of a family member, the approach of a higher ranking female, or the arrival of a musth male, and not imagine that they feel very strong emotions which could best be described by words such as joy, happiness, love, feelings of friendship, exuberance, amusement, pleasure, compassion, relief, and respect.[33]

Cynthia Moss has "no doubt" greeting elephants "are experiencing very intense emotions."[34] And she has "no doubt even in my most scientifically rigorous moments that the elephants are experiencing joy when they find each other again. It may not be similar to human joy, or even comparable, but it is elephantine joy."[35] Wildlife biologist-turned-writer and elephant-watcher Douglas Chadwick is "convinced that elephants experience delight, and I do not think that it would take any observer of them long to reach the same conclusion."[36]

The emotional cords that bind elephant families are acknowledged even by professional elephant exterminators who kill for economic, alleged environmental, even aesthetic reasons (some park managers don't like to have too many elephants about). They may pump .458 and .308 semiautomatic bullets into the spines and heads of terrified elephants herded to the killing grounds with low-flying light planes or helicopters, or shoot scoline from helicopters into elephant muscles, leaving them paralyzed, but aware that men are firing bullets into their family's brains. Some slay entire families to avoid leaving emotionally devastated survivors. The possible exception is the babies. Exterminators may tie the youngsters to their dead mothers while they are being butchered, then sell them to game farms, circuses, or zoos.[37] But elephant relationships stretch so far beyond mere family that, short of annihilation of an entire herd, there are always survivors, and lots of them.[38]

Surviving babies are scarred. Daphne Sheldrick, who rears elephant orphans in the Nairobi National Park, wrote to Poole that grief is a major barrier to an orphaned baby's survival. This may explain why, at Amboseli, when mothers die, 70 percent of calves between ages two and five, and 50 percent between five and ten, die within two years.[39] During their first nights at Sheldrick's orphanage, babies may wake up screaming. She assigns them human sleeping companions to help soothe their fears. Collins Ajouk, a soft-spoken University of Kenya accounting student, volunteers at Sheldrick's orphanage. Standing in the midst of a gang of lively babies who were kicking a soccer ball in the heat and rolling in the red Kenyan mud, he said that when the babies arrive, they often lash out at every human. They grieve, they cry, and it can take them some time before they begin to trust.[40]

Elephant Memory and Learning

Elephant memory has passed into legend, exemplified by H. H. Munro's epigram (writing as Saki): "Women and elephants never forget an injury." Poole says they remember humans unseen for a dozen years. Others claim that after many decades, elephants remember people who studied them.[41] Four centuries ago, Francis Bacon called revenge "wild justice," to be strictly checked by human law.[42] It is the only justice available to elephants. They seem to nurse grudges and take the coldest of revenges.[43] One Indian elephant veterinarian told Douglas Chadwick, "If a mahout is cruel, one day, sooner or later, perhaps a decade or more may pass, the elephant will try to kill him."[44]

Matriarchs' memories are a storehouse of critical social information. Adult females may remember the calls of a hundred elephants, depending upon how well they know the caller. The ability of older matriarchs to recall accurately and discriminate between familiar and unfamiliar elephants is most likely why families with a matriarch older than thirty-five are thousands of times more likely to bunch in self-defense upon hearing the call of an unfamiliar elephant than are families whose matriarch is younger than thirty-five.[45]

Due to the great difficulty of keeping elephants in humane captivity and of testing them, few formal cognitive tests have been attempted. Elephants are so immensely strong that a swipe of the trunk can kill. They are expensive to house and feed, and that's why most of what we know about elephant minds has emerged from decades of patient observations of wild elephants by Iain Douglas-Hamilton, Cynthia Moss, Joyce Poole, Katy Payne, and a handful of others.

A few souls have tried to administer formal cognitive tests to these giants; these suggest there is much to elephant memory and learning. Bernard Rensch, then director of the Munster Zoological Institute in Westphalia, West Germany, presented a series of cognitive tests to a five-year-old female Indian elephant in the early 1950s. He built a pair of wooden boxes, each somewhat larger than a hardback book, and inserted slots through which he could slide a cardboard card. The elephant could pick up the lids with her trunk. He began by inserting a card marked with a circle into the slot atop one box and a second card marked with a cross into the slot over the other box. Within the box marked by the cross he placed a piece of bread. She figured that out and eventually learned to differentiate between twenty pairs of symbols, such as a bird paired with an insect, black coupled with white, circle and semicircle, wider and narrower stripes, and six pairs of musical notes. A year later, she remembered written symbols and sounds with a high degree of accuracy.[46]

In the 1960s, Leslie Squier began a series of star-crossed tests with Rosy, Tuy Hoa, and Belle, three Indian elephants *(Elephas maximus),* at what is now the Oregon Zoo. The two elephant species separated less than 5 million years ago.[47] Though formal comparative research is lacking, Moss thinks their mentalities appear so similar she would comfortably apply cognition results from one species to the other.[48] Their chief difference, she says, lies in another area, temperament, and she compares the two species to horses; like a thoroughbred, the African elephant is more skittish and high-strung than her more quarter-horse-like Asian cousin.[49]

The three Oregon Zoo elephants could choose between light and dark by reaching through the bars of their cage and punching a plexiglass disk with their trunks. Within a few sessions, they were responding correctly more than 90 percent of the time. But the zoo administration stopped Squier's testing. Then his data were destroyed in a fire. Eight years later, Squier, with colleagues, retested the three elephants with the same equipment. Tuy Hoa immediately began pressing the disk, getting eighteen right out of her first twenty choices. She took just forty-three trials and six minutes to get twenty right in a row.[50] But Belle was still struggling after 2,863 trials over almost twelve hours, Rosy after 1,240 trials over three and a half hours. After scratching their heads, the investigators decided to check their vision. Sure enough, memory was not the problem. The two elephants couldn't see.[51]

As late as 1991, Charles Hyatt bemoaned that no formal research on African elephant learning had even been undertaken.[52] He replicated some of Rensch's work with two African elephants, Zambesi and Starlet, both orphaned through culling operations, and purchased by Zoo Atlanta.[53] Douglas Chadwick reminds us how difficult Rensch, and now Hyatt, had made the symbols the elephants had to learn. They

> were not simple dots and squares . . . but complex symbols. Very complex. "It is also interesting to note," Hyatt and his coworkers wrote in their report "that in both elephant studies, even after thousands of trials, the human experimenters relied heavily on written notes to identify the correct stimuli." In other words, the humans couldn't always remember which cards went with which.[54]

Zambesi remembered sixteen patterns out of twenty over eight months later; Rensch's subject remembered thirteen of twenty after a year.[55] Hyatt suspected that because vision isn't their best sense, the elephants might have learned even better if experimenters had tested them through their sense of hearing or smell. [56]

The most comprehensive work has been done by Robert Dale and Melissa Shyan, with David Hagan, on the spatial memories of five wild-born female African elephants, Tombi (16), Kubwa (17), Sophi (26), Ivory (11), and Cita (25), at the Indianapolis Zoo. They learned to choose from an array of two, four, then eight large food-containing pots and remember which pots they had visited so they would not revisit an empty one.[57] Tested more than six months later, they performed just as accurately.[58] Three Asian elephants, Ambiki (45), Shanthi (17), and Toni (27), and a sixth African elephant, Nancy (39), were just as successful, even when Dale, Shyan, and Hagan tried to disrupt visual or olfactory clues to which pots contained food.[59] The investigators noticed that Cita and Ivory shook their heads in apparent frustration or annoyance when they found an empty pot. (Recall the frustrated and displeased reactions of Alex and Griffin when they expected to find a tasty nut beneath a cup and found a less palatable nugget instead.)[60]

Elephant Self-Awareness

Few elephant self-awareness tests have been performed. In the late 1980s, Daniel Povinelli gave mirror self-recognition (MSR) tests to Shanthi (12) and Ambika (39), two Asian elephants at the National Zoo in Washington, D.C., using a mirror three and a half feet by eight feet. Both failed.[61] Neuroethologist Lesley Rogers skeptically eyed Povinelli's findings: large as the mirror was, it wasn't big compared to elephants. One also has to consider that elephants' eyes lie on the sides of their heads, which requires them to look mostly sideways. They just may not recognize themselves from their heads. Elephants may also recognize other elephants, and perhaps themselves, by what they hear, smell, feel, or touch, and not by what they see. Povinelli's test, Rogers argued, may indeed show that elephants cannot visually recognize themselves, but that's all.[62]

It turns out elephants may demonstrate MSR. Bertha, an Asian elephant, has performed at the Nugget Hotel in Sparks, Nevada, for forty of her forty-five years. Eight-year-old Angel joined her at age six months. In the late 1990s, psychologist Patricia Simonet erected a four-by-eight-foot mirror in the large yard in which the elephants spend time every day. After giving them two weeks to get used to the mirror, Simonet marked both elephants on their foreheads, and elsewhere, and videotaped their responses. Bertha noticed the paint within five minutes. Both used the mirror to investigate the marks on themselves as well as to locate objects Simonet had hidden that could be found only with the mirror.[63]

Nineteen hundred years ago, Aristotimus, one of Plutarch's dialoguers, was unsurprised that performing elephants learned movements that were complicated, varied, and difficult, even for humans. But he thought elephants' intelligence was truly revealed in their feelings and movements offstage, as in Rome when an elephant, thinking himself alone, was seen practicing his dance steps before taking the stage.[64] Joyce Poole thinks elephants have some sense of self.[65] One reason is their reactions to elephant dyings and deaths suggest some concept of death.[66] Though it's "probably the single strangest thing about them," Cynthia Moss isn't sure what's going on, though something is.[67] Elephant death clearly disturbs elephants. They become quiet and tense, and approach a body slowly.[68]

Elephants ignore nonelephant remains, except perhaps those they killed. Katy Payne relates the story Peter Ngande, the cook and general assistant at Moss's camp, told about a lion in another part of Kenya who leaped onto the back of an elephant cow.

> In a single motion the elephant reached her trunk over the lion's body, grabbed him by the tail, ripped him off, and, using the tail as a handle, slammed him onto the ground repeatedly until he was dead. The elephants then broke branches from some nearby bushes, and covered the dead lion with them—a sort of burial—before they walked off.[69]

Most striking is their interest in elephant bones and tusks. In tense silence, they may sniff, taste, caress, and hold them, run their trunks into every crevice of the skull, move bones, or carry them away. They may bury a body with branches, palm fronds they break off, ground vegetation, or dirt.[70] They have been known to pass ivory and bones one to another around the herd.[71]

When the EBs encountered the carcass of a young elephant in a clearing,

> [t]hey smelled and felt the carcass and began to kick at the ground around it, digging up the dirt and putting it on the body. A few others broke branches and palm fronds and brought them back and placed them on the carcass. At that point . . . I think if they had not been disturbed (by a swooping plane) they would have nearly buried the body.[72]

Zoologist Karen McComb replayed the call of a dead Amboseli elephant to her family. They immediately moved toward the speaker. But Moss and Poole rejected the proposal of a scientist to research this further, believing it would, at minimum, cause confusion. At worst, it might reignite the family's grief.[73] Echo once investigated the bones of Emily, perhaps her sister, dead six months. She felt the lower jaw and ran her trunk along the teeth, which is where elephants often place their trunks when greeting.[74] After Tuskless, matriarch of the TA family, was killed, Moss sorrowfully collected her skull and added her lower jaw to the ring of jaws around her tent. When Tuskless's sister, Tillie, entered Moss's camp with her calves for the first time since the matriarch's death, they went straight to her jaw and handled her skull each in turn.[75] Moss suspects this special attraction to lower jaws, skulls, and tusks may mean elephants are trying to recognize the dead individual.[76]

Marc Hauser writes "to have an understanding of death is to have specific beliefs about what it means to be dead," and to understand death as something that happens to the living depends upon having self-awareness. He agrees an understanding of death explains the elephants' behavior but argues that other explanations can't yet be ruled out.[77]

Elephant Problem Solving and Tool Use

Anecdotes about elephant tool use and problem solving abound. Douglas Chadwick reported zoo captives turned on water faucets and unfastened shackles. At the Phoenix Zoo, they opened water valves. After zoo staff bolted steel plates over the valves, the elephants broke the nuts with rocks and opened them again. When Chadwick visited Phoenix, a handler hosing a stall jokingly said to an onlooking elephant, "Rafiki, you're in the way. Why don't you go outside and chain yourself?" She did.[78]

At Marine World, Tava used a log to pry open a retaining wall. When a strip of spikes replaced the wall, the elephants bridged the spikes with other logs. Elephants throw tires onto tree branches to bend them to where they can eat the leaves. Wild elephants disable electric fences by dropping or throwing heavy rocks or logs onto them. One wild Asian elephant, stuck in mud, broke tree branches and placed them under her feet, which stabilized her until she could be rescued.[79] Amboseli elephants use sticks to scrape ticks from their bodies and palm fronds to swat flies and rub their eyes or ears to stop itches.[80]

Using trial and error, Rensch's Indian elephant took 330 trials over several days to puzzle out the first problem she faced, whether she should choose the box marked by the cross or the circle to get bread (Rensch and his colleagues helpfully shouted "*Nein*" when she chose incorrectly).[81] She learned to discriminate between the fourth pair of symbols in just ten trials.[82] In a final multiple-choice test, all twenty symbols were shown thirty times, 600 presentations. She kept the meanings straight and made few mistakes.

Rensch painted dots on cardboard cards in many configurations to test whether she could tell four dots from three.[83] No matter how the dots were painted, she succeeded. She learned to distinguish between pairs of musical notes, sometimes just a note apart. Then he taught her two brief melodies. No matter how the two patterns "low note, high note, low note" and "high note, low note, high note" were presented, she almost always distinguished them.[84]

Twenty years ago, Ben Beck introduced an oft-cited definition for tool use: using an unattached object to alter something else. A decade later, when describing elephant tool use, Suzanne Chevalier-Skolnikoff and Jo Liska thought Beck's definition a tad broader than the usual one, the use of a physical implement to attain some goal.[85] When they combed the literature, they found seventeen types of tool use by African elephants, eight by Asian elephants.[86] Those that met the stricter definition included reaching toward food with a stick, cleaning out an ear with grass, and brandishing branches at a vehicle. They also reported that an African park service built, then used, a road to cull elephants. Four times elephants broke off branches, carried them to the road, and piled them onto the road until it was finally closed.[87]

Observing three captive African elephants at Marine World Africa USA for thirteen hours, Chevalier-Skolnikoff and Liska saw nineteen kinds of tool use (using Beck's looser definition), at the rate of twenty-two acts and nine bouts (groups of acts) per hour. Using the commoner tighter definition, they observed five acts, such as throwing objects at birds eating their food and using vegetation to sweep the ground before they placed fruit on it, and one bout each hour.[88] In Kenya for six weeks, using Beck's definition, they witnessed nine types of elephant tool use, at the rate of one and a half acts and half a bout of tool use per hour, or two types using the narrower definition, at .15 acts and .08 bouts an hour, which included using the trunk to swat one's body with vegetation and scratching one's body with vegetation held in the trunk.[89]

In southern Africa, elephants have been seen digging wells sometimes four feet deep. Mothers try to keep their babies from interfering. Katy Payne tells how one was saddled with a calf who refused to be dissuaded, even by kicks. He kept breaking down the walls of the well.

> Then she drew him aside to an empty spot about ten meters from the well and scuffed open a deep, wide, cool, damp, wallow. She pushed him into the wallow and scooped damp sand over him until he was engaged in the wallow as his own playful project. Then she withdrew to her well, and finished it, free of distraction.[90]

In the 1950s, James Gordan was a Tsetse field officer in South Africa. Following an elephant path, he came upon a dry place where elephants had scraped away some eighteen inches of sand. When he was game warden in the Wanki National Park in what was then Rhodesia, he saw holes that elephants dug while searching for water. He stopped to see whether an elephant had actually found water. Excavating to one side of the hole, he touched something hard, reached in, and pulled out a long, wide piece of bark that an elephant had chewed into a ball, then used to plug the hole. He could not reach the bottom with his arm. But a dipped stick emerged dripping. And a short distance away stood the baobob tree from which the elephant had stripped the bark.[91]

Elephant Communication

Investigation into how elephants communicate is in its infancy. Katy Payne's discovery in the 1980s that elephants communicate by low-frequency rumbles (11 Hz) cheered many a veteran elephant observer. For years, they had watched entire families suddenly freeze or alter course so they wouldn't meet, even when out of each other's sight. Iain Douglas-Hamilton told Douglas Chadwick that no one had a clue how they did it. "We didn't mention ESP openly," he said, "but I can tell you that some of us were ready to entertain the idea that these animals were sending bloody mind waves to each other."[92]

They not only rumble, but roar, bellow, groan, cry, snort, trumpet, and scream, sometimes at a deafening 117 decibels (a jet on a runway emits about 120 decibels).[93] Joyce Poole has catalogued seventy-five Amboseli elephant calls, noting the context in which each occurs and the quality of the sound. Some locate, contact, call to, greet, or reassure family and bond group members. Others indicate a caller's emotional or physical needs or desires or inform, instruct, and negotiate with other elephants. She suspects the "let's go rumble," which a female gives while standing in the direction she wants to head, then repeats until others follow, and the "attack rumble," which can place an entire family on the offensive, and perhaps other communications, function symbolically, like a word.[94] She has even documented "discussion rumbles" that move back and forth, sometimes for twenty minutes, with the cadence of a sluggish conversation.[95]

Elephant Play, Pretense, and Deception

Approached by Joshua, a teenager from the JA family, Joyce Poole tossed him her flip-flop shoe. He investigated, scratched with it, chewed it, stepped on it, vaulted it in the air. Then he tossed it back to Poole. She threw it back. So did he. When she interrupted their game by glancing away, he brained her with a wildebeest bone. They were, she writes, just "two species out on the plains playing catch." Other elephants have tossed sticks, stones, even a film box at her, sometimes in play, sometimes in pique. They do the same with each other.[96]

Elephants take play seriously. One Kenyan morning at Tsavo National Park, Chadwick watched unsuspecting elephant youngsters at play: "Small juveniles sparred with bigger ones, who dropped to their knees to make the contests more even. In one mock battle each flourished a broken branch high overhead in its trunk like a flag."[97]

Moss has seen the same thing, a large bull lying in an upright position on his knees with his rear legs splayed behind him, sparring with a much smaller opponent.[98] She often finds the youngest EBs playing in a way any human parent will recognize: Eric, Emo, and Edwina, half-submerged in a deep pool near the Enkongo Narok swamp, shoving and climbing over each other, racing through the water, smacking it with their trunks, submerging themselves, completely or to the tips of their trunks. One-year-old Edgar kicks dried dung about like a soccer ball. Other youngsters chase wildebeests or baboons, clamber atop older calves, trunk-wrestle, throw sticks, and play-trumpet. Amboseli youngsters did so many silly things that Moss simply created a formal scientific category for what they were doing and called it "being silly."

Not just youngsters play. So do older females. Echo may play with a palm frond or play-trumpet and play-charge Moss's car. Entire families "floppy run," head down, ears and trunk dangling loose, play-trumpeting. After watching one extended bout of floppy-running, play-trumpeting, and mock-attacking, a half-exasperated Moss wrote in her field notes: "How can one do a serious study of animals that behave this way!"[99]

There is some evidence elephants engage in pretense, deception, and teaching. They appear to attack imaginary opponents by charging, throwing branches and sticks, or trumpeting loudly.[100] Judy, a captive Asian elephant at Marine World, would bunch her chain, stand on it, and act as if she couldn't reach her hay. If a trainer was foolish enough to push the hay closer, Judy would step forward and smack him with her trunk. Three trainers eventually went down.[101] One handler at the Phoenix Zoo claimed one elephant who had figured out how to remove his shackles when alone would quickly reshackle himself when he saw someone coming.[102]

When Echo's Ely was born crippled, for the first few days of his life, his sister, Enid, appeared to try to show him how to walk by slowly moving past him, again and again, all the while looking over her shoulder to see if he was paying attention.[103] When a daughter enters her first estrus, she finds herself the object of the attention of a great many large bulls. Confused, she may run away. Experienced females attach themselves to the highest-ranking musth bull, who will chase competitors away. A mother will often lead her daughter to that bull, stand with her and model the movements her daughter is supposed to make. Poole says the behavior of mother and daughter can be so similar that it's hard to figure out who's in estrus.[104]

Echo

As I sat beside Cynthia Moss watching elephant families living their complicated, close-knit lives on the dry Amboseli plains, playing, greeting, touching, trunk-swinging, rumbling, eating, drinking, I couldn't help thinking of another scene.

It was of a depressing spectacle that occurred in Madison Square Garden on April 9, 1942. Fifty elephants appeared with fifty female dancers. They were forced to "dance" to a ballet called the "Circus Polka: Composed for a Young Elephant." The dance floor was in the Ringling Brothers and Barnum & Bailey Combined Circus. It lasted four minutes. Music was by Igor Stravinsky. The choreographer was George Balanchine. Balanchine's wife, Vera Zorina, was the main dancer. Each woman wore fluffy pink. Each elephant wore a tutu, a headdress, and a large set of earrings. They were forced to repeat this 425 times over the coming year.[105]

Douglas Chadwick wrote:

> If I learned anything from my time among the elephants, it is the extent to which we are kin. The warmth of their families makes me feel warm. Their capacity for delight gives me joy. Their ability to learn and understand things is a continuing revelation for me. If a person can't see these qualities when looking at elephants, it can only be because he or she doesn't want to.[106]

Did I find Echo in camp? Not that time.
"Very Echo-like," Moss said.

Does Echo Have Practical Autonomy?

With Marbury and the Atlantic bottle-nosed dolphins, Ake and Phoenix, Echo, and I last shared a common ancestor about 90 million years ago. The world's most experienced elephant-watchers are certain African elephants intensely experience a wide variety of emotions: joy, happiness, friendship, exuberance, delight, amusement, pleasure, and compassion. They have long and powerful memories, and there is anecdotal evidence that they expect or believe, as revealed in their reactions to the elephant death.

Richard Leakey relates how Joyce Poole drove him into the midst of an Amboseli elephant family the way Cynthia Moss drove me, stopping and proceeding to chat about their feelings and behavior.[107] For one reviewer of Leakey's book, "[t]he point is not simply that elephants are 'intelligent,' in a maze-solving way. It's their awareness that startles: their sensitivity and family feeling. To poach them looks very much like murder."[108] There are few African elephants outside of Africa. They number only in the thousands and have little economic value. Although I have not seen surveys of attitudes toward elephants, they appear beloved. Thus there seem to be few physical, economic, or elephant-specific psychological barriers to placing Echo where the scientific evidence indicates she should be placed, in Category Two. African elephant tool using, problem solving, and communication, along with sparse but suggestive evidence they pretend, deceive, and teach, certainly reveal some ability to desire and act intentionally. But how high up in Category Two should Echo be placed? One important reason for placing her higher, rather than lower, is because Indian elephants, at least, appear to have passed a mirror self-recognition test, it seems likely that Echo would too. But no African elephant has actually yet passed, and until one does, Echo's success remains just a probability. However, she probably has an ecological and interpersonal self, as well as James's "I-Self."

I therefore assign Echo an autonomy value of 0.75, above Marbury and just below Alex. The primary reason she tops Marbury is the likelihood she possesses mirror self-recognition. If an African elephant clearly passes an MSR test, Echo's cognitive value should be moved to 0.80, even higher. If MSR cannot be demonstrated, her autonomy value should drop to Marbury's 0.68, or even lower, for dogs demonstrate limited Substage 6 abilities, for which African elephants have not been tested. On the other hand, she lacks the extraordinary and sophisticated language abilities that Alex possesses, with all that implies. There is sparse evidence she can use symbols, as Alex can. But she has never demonstrated Substage 6 object permanence, as Alex has. Nor has she shown she can imitate either at the action level or program level, as Alex can.

TEN

Chantek

Lyn Miles drove through a staff entrance at Zoo Atlanta. She stopped and pointed toward an outside enclosure. One of three orangutans stirred from a hammock and shuffled to the edge of the enclosure, the way an astronaut might after an extended flight. I knew orangutans are the world's largest arboreal animal, but I didn't see any trees inside his enclosure. His enormous cheek pads made him appear immense.

"How much does he weigh?" I asked.

"Three hundred pounds," Miles said.

"His head . . . "

"When they gain too much weight, their cheek pads balloon. He's on a diet now."

"YOU." The orangutan gestured with his right hand. Miles began a simultaneous translation. "YOU, YOU, YOU, YOU, LYN."

When Siena was two, she called Christopher and her mother "You." Miles lifted her hands high above the steering wheel.

"SECRET," she signed.

"WHERE?" asked Chantek.

"OVER THERE," Miles signed.

Chantek moved his fingers.

"He's asking for fruit," Miles said. "He knows he's not allowed to have fruit. I'm going to tell him about you."

"NEW FRIEND. GOOD MAN BRING FOOD." She turned toward me. "He'll like that." She clenched her fist and placed it on her forehead. "STEVE." She stared at Chantek. "That will be your name sign. An "S" on the forehead."

"STEVE," Chantek signed. "STEVE."

"Wait here while I get the food out of the trunk and bring it over," Miles said to me.

She carried the groceries toward the enclosure. At her signal, I walked toward them. Miles was laying out grapes and oranges. As she walked, she said Chantek often sends her on errands to buy stuff, especially Naya, Canadian bottled water. Children's puzzles—some the same as Siena's and Christopher's—were stacked beside her. She handed me latex surgical gloves.

"It's the rules."

She was apologetic. She gave Chantek a puzzle from "Lights, Camera, Interaction." Each piece was cut in the shape of a different motor vehicle. He dumped the pieces on the cement, then placed the puzzle board on his lap. He picked each piece up and began turning it around until it fit a hole. Miles held out an orange. He placed a clenched fist on his forehead.

"STEVE."

"He wants you to give it to him."

I offered it. He slowly smothered it in huge puffy fingers and withdrew it into the cage. Miles began to play "Simon Says."

"Simon says 'clap,'" she said.

Chantek clapped.

"Simon says 'pat head.'"

He patted his head.

"Simon says 'pat right shoulder with left hand.'"

He did it.

Whenever Chantek made a sloppy sign, Miles said, "Sign better." He would try again. If the sign was still slipshod, she reminded him, "Sign better." And he would. After an hour, Miles held up a grape.

"STEVE," Chantek signed.

Miles handed me the grape. I passed it through the bars. Chantek grasped it between one Brobdingnagian thumb and index finger, then "accidentally" squeezed the tip of my index finger. It felt like a vise (I write "accidentally" because more than one scientist familiar with orangutans suspects Chantek saw me as a mark from whom he could steal a glove to use to trade for goodies).[1] Instinctively I jerked away. But he caught the tip of my finger.

"Let go!" Miles said sharply. "Let go!!"

He did.[2]

Chantek is the most studied orangutan, having been formally observed for more than 35,000 hours.[3] His enculturation began within a few months of birth. Chantek was born to Datu and Kampong, two wild-caught orangutans held captive at the Yerkes Regional Primate Research Center. When Chantek was nine months, Miles, who wished to study the linguistic and cognitive development of an enculturated orangutan, installed him in a five-room 12-foot by 40-foot trailer on the campus of the University of Tennessee in Chattanooga. There she immersed him in a human cultural and social environment, and there he was educated by a cadre of caregivers/companions until, suddenly, at the age of nine, he was forced by circumstances to return to Yerkes.[4] Yerkes passed him to Zoo Atlanta. Miles has been fighting to regain complete access to him ever since. She says she feels his loss as keenly as if he were her son.

Taxonomy, Genes, and Brains

"Orangutan" comes from "orang" and "hutan," meaning "person of the forest" in Malay.[5] Mild and relaxed, the orangutan is the Zeppo to the chimpanzee's frenetic, emotional Groucho, Abbot to their Costello.[6] Taxonomists traditionally place them within the family *Pongidae,* along with chimpanzees and gorillas.[7] Traditional taxonomy is subjective, and it's hard to envision how evolutionarily related animals are in objective ways. One might try to pinpoint the degree to which we share DNA, as I did in *Rattling the Cage,* for chimpanzees, bonobos, and humans. In the 1980s, Charles Sibley and Jon Alquist found that human and chimpanzee DNA was 98.4 percent identical; human and gorilla, 97.7 percent; and human/orangutan, 97.4 percent. Many scientists are sure that some DNA is "junk" and that the working DNA for humans and chimpanzees is perhaps 99.5 percent the same.[8]

The ancestors of orangutans probably split from the common ancestor of humans and the other great apes 13 or 14 million years ago.[9] That means you need to hold your breath for a little less than five seconds, not much longer than the two seconds needed for chimpanzees. Chantek is obviously a close relative.

Orangutans spend much of their lives in trees and are easily the most solitary ape. As late as 10,000 years ago, they lived throughout Southeast Asia, from Thailand to Beijing, Taiwan to Java. Today, not many more than 10,000 live in scattered populations confined to the rain forest of the Malay Archipelago islands of Borneo and Sumatra in Indonesia. Unless extraordinary measures are taken, they will soon become extinct in the wild. Orangutans give birth about once every eight to ten years. Birute Galdikas, who studied orangutans the way Jane Goodall studied chimpanzees and Dian Fossey gorillas, says that relationships between mothers and infants are "extremely intense."[10] The period of immaturity in the wild is the longest of the great apes, between nine and twelve years, perhaps a quarter of their life span.[11] Infants may nurse for six years and weaned juveniles may stay with their mothers for several more years.[12] Females may not reach adulthood until age fourteen to sixteen, males not until age nineteen or twenty, though fifteen is more usual.[13]

Mental Representation

Comparative psychologist Anne Russon usually observes orangutans in the jungles of Borneo, but she'll watch them anywhere. While studying eight at the Toronto Zoo, she noticed they sometimes play with their eyes closed or covered. If they bump into something or miss what they're grabbing for, they might "cheat" by blindly groping about or peeking for a few seconds to get their bearings. Sometimes they stare at a location, orient in that direction, cover their eyes, and set out, ending up precisely there, never backtracking and only occasionally peeking. Ramai, an adolescent female, once swooped blindly through a snarl of bars, poles, and ropes and crossed three times between unconnected climbing structures, all within fifteen seconds. She could obviously "see" with her mind's eye.[14]

In planning goals and routes, wild orangutans show they, too, mentally represent. They love wild durian fruit. But these are few and far between; only one durian tree may grow in every one and a half square kilometers of forest.[15] Yet orangutans find them, usually by direct routes through the jungle.[16] They may return to the same tree every time it flowers—usually once every two or three years—and accelerate as they approach a familiar source of food or water, long before they could have sensed it.[17]

Until recently, field researchers spotted orangutans using only simple tools, a leaf to swat flies or wipe feces from their hair, a leafy branch as an umbrella, perhaps a stick as a back-scratcher.[18] That has changed. Now they're seen stripping leaves and twigs from branches to make tool kits of two kinds: insect tools and fruit tools.[19] Longer and wider tools, which they modify as they use them, are used to hammer, poke, and probe tree holes in search of insects and honey. Seeds are pried from fruits, whose stout husks nearly defy hand-breaking, with thinner shorter tools. Cracks in the husk of one fruit widen with ripening.[20] As these cracks expand, orangutans insert wider tools to force the husk. They manufacture tools from branches high in the trees.[21] Orangutans may replace tools that become useless or fall by popping the tips of new ones into their mouths and carrying them to other feeding sites.[22] Orangutans must be able to maintain mental representations of a tool's physical properties long enough to fashion raw materials into new tools or to alter existing ones.[23]

Captive orangutans or those being rehabilitated for return to the wild after living with humans (rehabilitants) are "unparalleled" tool users, superior even to chimpanzees.[24] Russon calls them "mechanical geniuses," and it's in the context of their tool use that we can see their mental representation most clearly.[25] They use tools to build swings. They make tools by fastening sticks together to rake in food.[26] Russon says, if given a chance, some orangutans prefer to eat messy foods with spoons on plates and drink river water from bowls or mugs.[27] They sleep in hammocks they erected themselves and untie triple and even quadruple-tied knots to steal canoes for downstream joyrides.[28] Birute Galdikas watched one orangutan, Supinah, break eggs into a cup, add flour, and blend the ingredients. If she had cooked the batter, she would have made pancakes.[29] In *Rattling the Cage,* we learned the bonobo, Kanzi, could create stone tools.[30] The inspiration for allowing him to try was the 1972 work of Abang, a captive orangutan. He learned in a few hours to "knapp" a flake from a preshaped flint core researchers had tied to a plank, then used the flake to cut a cord and open a box containing food.[31] In doing so, he used a "meta-tool," a tool to make another tool.

Some orangutan tool use appears insightful. Remember, insight is often synonymous with thinking, as it involves mental manipulations that allow one mentally to see a problem's answer without engaging in extensive trial and error; it often allows an animal to attain a goal efficiently and safely.[32] Manufacturing seed-extraction tools before they are needed involves foresight. They must have some idea of what sort of tool they want to make before they make it, sometimes even before it's needed, as when Galdikas saw an orangutan drag logs to a stream, fashion a bridge, then cross.[33] Elizabeth Fox says they make a pretty good tool on the first try without going through a process of trial and error, even if it's not perfect.[34] Constructing tool kits requires orangutans to maintain multiple representations of tools and relate each tool to a potential use.[35] Russon says captives who blind themselves to play often peek or grope just before they reach a difficult juncture, that suggests they use the stored representation to 'see ahead' to the upcoming difficulty."[36]

Miles tells Chantek to "SIGN BETTER" when his signing gets sloppy, and he does. His teachers have always asked him to "SIGN

BETTER" and he does. He might even use one hand to mold the other into the correct shape. Or sign with his feet! That means he must have some mental image of what the correct sign looks like.[37] Chantek has completed sequences of tool use involving twenty-two steps. He obtained wire cutters from a box he opened with a key. Then he used the cutters to release a screwdriver needed to unscrew two pieces of wood.[38] Another orangutan, handed a piece of wood and a clear plastic tube into which a treat had been wedged, repeatedly tried to scoop out the treat through the plastic. After giving up, he ambled to his nearby sleeping blanket and manipulated it with the wood, while occasionally peeking toward the tube. Suddenly he returned to the tube, rammed the wood down its length, and knocked out the treat.[39] For comparative psychologist Gisela Kaplan and ethologist Lesley J. Rogers, if an animal confronted with a problem appears to ponder, then "more or less, goes deliberately to the solution," we can reasonably assume insight.[40]

When Josep Call recently compared the object permanence abilities of orangutans, chimpanzees, and nineteen- to twenty-five-month-old human children, he found not only that all three generally solved even the invisible displacement problems required for Piagetian Substage 6 competence, but performed at just about the same levels. He also showed that any problems all three had with the more complex complems stemmed not from a memory deficit but from an inability to inhibit certain behavior.[41]

Logical and Mathematical Abilities of Orangutans

We know relatively little about whether orangutans can count, mentally combine, or choose the larger of two quantities. Counting is complicated. If I ask you to count ten candies I place on a desk, you must understand each candy is to be counted as one and just once, that one comes before two, two before three, nine before ten, that ten refers to the total number of candies, and it doesn't matter in what order you count the candies. Until the age of four or five, humans usually can't count. But they can subitize (use a fast unthinking perceptual tally of items) to about four.[42]

Chantek and two mother-reared captive orangutans, Teriang and Solok, were able to select the larger of two quantities of breakfast cereal when presented with two dishes simultaneously; and again when the two dishes were covered; and when one dish, then the other, was uncovered, then covered again.[43] At minimum, they remembered and mentally compared two different quantities. On another test, at the very least, they remembered and mentally compared three different quantities, and Solok appeared to engage in some process of mental combining.[44]

It turns out orangutans in general, and Chantek in particular, are pretty good at conservation of quantities, not as good as eight-year-old humans, but competent as most six-year-olds and many seven-year-olds. The conservation of liquid quantities occurs when an animal understands the amount of a liquid doesn't change when it's poured into different-sized containers.

When I wrote this, Christopher was two years old. I showed him two identical glasses of water.

"Christopher," I said, "which glass contains more water?"

"Deck," he replied, and pointed over my shoulder. It was warm and sunny and he wanted me to open the sliders onto our deck.

"This first," I said. "Which glass contains more water?"

"Deck!" Point. Point. I gave up, opened the slider, and he ran out. I tried again yesterday.

"Which glass contains more water?"

"Deck."

"No, which glass . . . "

"Deck! Deck!" He started to whine the way he does when I'm not moving fast enough. I could never be a child development psychologist. Mercifully, I woke this morning to a pelting rain.

"Which glass?" I asked.

He gloomily pressed his nose against the slider. "Same!" I triumphantly emptied one glass into a thinner, taller glass and asked again.

"Big!" he yelled.

In a year or two, he'll probably make the same choice. But he'll be able to tell me why. He may say he chose the bigger glass or the taller. Piaget thought children of five or six "center," that is, concentrate upon some single characteristic and have difficulty holding two things in mind.[45] Because they can't simultaneously focus on the height and width of a glass, their reasoning ability is impaired. Piaget said, in three or four years, Christopher will enter a transitional stage of "partial conservation." He'll realize he needs to think about more than just one thing, and might even be able to answer my question. But he'll be stymied when one of the smaller glasses is then poured into two or more even smaller cups, because he'll still be unable to keep track of everything he needs to think about. It will take him a good six years before he'll be able to look beyond appearances and weigh all the evidence. Then, of course, he'll have less patience for my questions. Undaunted, I will press on.

"Which glass contains more water?"

"Duh."

"No. Really. Which one?"

"I'm late for soccer, Dad!"

"Humor me, son. Which one?" (Did I mention I could never be a child development psychologist?)

Now able to "truly conserve" liquid volume, he'll get around to answering. "The same. Okay? Let's go!"[46]

It's not easy to figure out when orangutans think two glasses hold equal volumes of water. But researchers Josep Call and Philippe Rochat took advantage of their adeptness at choosing the larger of two volumes. They tested four orangutans, including Chantek, and ten human children ranging in ages from six years and six months to eight years and eight months by offering them a choice of drinks and allowing them to choose. After orangutans and children pointed to the container with the larger volume, Rochat and Call poured the liquid into two smaller containers, each shaped differently from the larger container and from each other. All four orangutans showed partial conservation by consistently pointing to the smaller container with the larger volume. Most of the children failed. Two eight-year-olds flunked all four tests and three fared more poorly than did the most incompetent orangutan.[47]

Rochat and Call suspect the orangutans responded based upon their estimate of the volume, irrespective of the container's shape, or by keeping track of the amount poured, or by detecting which container held the larger amount of liquid in the first instance and then keeping track of it.[48] They thought the failures revealed the orangutans' "remarkably creative" minds; they were "active problem-solvers, using a variety of strategies to solve novel problems . . . of a level comparable to most 7–8 year old children."[49] Chantek in particular flexibly used all three strategies to solve the problem, without trial and error, even though he was never given feedback on whether he was getting it right.[50]

Orangutan Self-Awareness

At ThinkTank, in Washington, D.C.'s National Zoo, Benjamin Beck, who twenty years ago defined a tool user as I mentioned in Echo's chapter, and his colleague, Rob Shumaker, perform orangutan cognition research in public view as part of the zoo's Orangutan Language Project. None of the orangutans are human enculturated; all are well past infancy, some even adults. Five of ThinkTank's six adult orangutans tested use a mirror to explore parts of their bodies they could not see without it. Three have passed a formal mirror self-recognition (MSR) test.[51] Not content with touching the mark, Bonnie gazed at it for a few seconds, then got water to try to wash it off and rechecked her reflection![52] Chantek passed first at the age of twenty-five months but did not reliably pass again for another sixteen months.[53] His MSR became more elaborate between the ages of four and eight. Miles became accustomed to Chantek's using the bathroom mirror to inspect and turn his face from side to side; to peer into his mouth; to groom his face, shoulders, and chest; and to examine a swollen eye and a cut lip.

He often named his mirror image while conversing with others or by signing to himself. He made faces in the mirror. At six, he received a pair of sunglasses, took them to the mirror, and modeled them from a variety of angles. Just shy of eight, he used the mirror to help him curl his eyelashes with an eyelash curler the way a caregiver curled hers.[54]

Joint Attention

When Siena ambles into the kitchen, looks at me, and points to a carrot on a high counter, she communicates. Her communication is "imperative"; she wants help to attain a goal.[55] But some animals sometimes just want to share the world, never intending to accomplish anything more than focusing joint attention upon some thing, some one, some act. Often it's done with gazes. Alternating eye gazes usually indicates intentional communication and a recognition that another is seeing what you see.[56] We often point to share. When Christopher puts a colander on his head and declares he's a fireman, Siena giggles and points.[57] That is "declarative" communication.

Chantek is the only orangutan whose pointing has been systematically studied. As an infant, he "pointed" with his whole body, leaning his weight and orienting his body and head in the direction he wanted to be carried, while looking into the face of whoever was holding him. At seventeen months, Miles began to teach him "GO," forming the letter "G" with one hand and thrusting it in the intended direction. Chantek first used "GO" to ask his caregivers to move him about. Then he began to point "GO" in specific directions.

Four months later, Miles taught Chantek a pointing game. Caregivers would ask, "Where is your nose?" Where are your eyes?" and point to his nose, his eyes, and repeat with other body parts, a game I endlessly played with Siena and Christopher during their second year. As they did, Chantek imitated their pointing upon his body. At twenty-six months, he began pointing to body parts he wanted tickled. When asked, "Where you want tickle?" he pointed. The next month, he spontaneously brought objects to someone's attention by pointing. And he would point to an object when asked "WHERE" it was. At twenty-nine months, he began to combine his signs for "GO" and "POINT" to indicate exactly where he wanted to go.[58]

Seven years later, Josep Call and Michael Tomasello drafted Chantek into a pointing study. He pointed to a rake that a human used to gather food for him; he did it quickly and without specific training. Also, he knew what humans meant when they pointed. When food was placed into one of three containers, he understood he could find it in the one to which a human pointed. He usually pointed only when a human faced him with open eyes, occasionally with eyes closed, almost never when the human had left the room or turned his back.[59] He showed intentional communication by alternating his gaze from the eyes of the human with whom he was communicating to the place he wanted them to look. Tomasello and Call concluded Chantek understood humans to be intentional agents, at least for pointing purposes.[60]

Chantek had a partner in these experiments. Raised in a nursery at the Yerkes Regional Primate Research Center, Puti didn't point. Call and Tomasello trained her to point at age twelve. But she never pointed spontaneously or grasped what pointing was all about; she failed to understand the difference between pointing toward a human whose eyes were open and one whose eyes were closed, although she rarely pointed when a human wasn't at least facing her. And she hardly ever alternated her gaze between a human and the place she wanted a human to see.[61] Call and Tomasello thought upbringing accounted for the difference between Chantak and Puti. However, an unenculturated orangutan tested in Japan grasped that pointing is intentional communication.[62] The ThinkTank orangutans also learned to point as adults; none were enculturated the way Chantek was. Rob Shumaker visited wild-born rehabilitant orangutans in Indonesia and reported that whether they did or did not have a great deal of human experience, "they point just fine."[63]

Gazing might be thought of as visual pointing. Some scientists think "visual referencing," alternating gaze between another and a goal, is necessary for intentional communication without language.[64] By about age two, wild orangutans alternate gazes.[65] The orangutan in Japan who understands that pointing is intentional communication follows a human gaze.[66] Chantek alternates his gaze between a human's eyes and what he wants that person to look at even outside the realm of pointing.[67] Dona, a fifteen-year-old orangutan captive in Spain, routinely pointed with her entire arm to the location of keys to open boxes containing food; she often accompanied her pointing with an alternating gaze. Although her goal may have been to obtain the hidden food, her pointing and gaze alternation weren't so much requests as proto-declarative pointing, intentionally executed to draw the attention of another to what she was pointing to.[68]

Orangutan Imitation

At ThinkTank, six orangutans divided themselves into three dominant and subordinate pairs, Azy and Tucker, Bonnie and Kiko, Indah and Iris.[69] Each dominant orangutan was designated the demonstrator; the subordinate orangutan, the observer.[70] Shumaker's plan was to teach each demonstrator how to match abstract symbols with apples and bananas, in six stages, over several months. The observer would watch. First, he taught the demonstrators how to point with their index fingers because orangutans usually extend an entire hand, or arm and hand. It only took a day or two.[71] Then he started to teach them symbols from a language board and asked the dominant to choose the correct symbol from various possibilities.[72] Shumaker noticed that on ten occasions the dominant orangutan was distracted and the subordinate stepped to the screen and pointed to a symbol with an index finger.

Except when Azy injured Tucker badly enough to put him out of the experiment for two months, all went smoothly. But Tucker's absence spoiled Shumaker's plan. Rather than complete the experiment with two pairs, Shumaker began testing the two observers, and Tucker when he was fit, to see how long it took them to reach the stage at which his or her partner was operating when the ambulance came for Tucker.

Shumaker found that the subordinate orangutans imitated their dominant partners. His most interesting evidence was that not only did the observing orangutans point with just their index fingers, which orangutans don't do, but nine out of ten times in which a subordinate interrupted, she got the answer right![73] And they stuck out their hands or lips for the same reward the dominant would have received. Yet subsequent testing made it clear they didn't understand what the symbols meant. They were imitating the dominant.

Asked to "do what I do," Chantek pointed to body parts and imitated actions, drawings, and sounds. Asked to touch his teeth, sound like Donald Duck, or draw a circle, he usually succeeded, almost always partially, and at the level of true imitation.[74] He developed imitation much as human children do, though delayed.[75] Christopher imitates my words. When young, Chantek learned to sign by having his hands molded. Later, he imitated signs made by others.[76] There would have been no way to teach Chantek by demonstration if he hadn't been able to imitate precisely.[77]

Chantek did not just mindlessly copy. Asked to make a noise with his hand, he might make it with his foot. When a caregiver blinked her eyes, Chantek might press his eyelids with his fingers.[78] And the "Simon Says" games he and Miles played were often real. Miles would announce the game, then sign "CHANTEK DO SAME." "Simon Says," she would say, "'Put your hands on your head.'" Orangutan hands would go to the head. "Simon Says, 'Put your hands on your eyes.'" Hands on the eyes. "Put your hands on your mouth." Hands on the mouth? "I didn't say 'Simon Says,' Chantek. You're out!" But he wasn't fooled every time. Eventually, Chantek learned to imitate what "Simon Says," not what Simon didn't say.[79]

Ten years later, after Chantek returned to Yerkes, Josep Call and Michael Tomasello tested whether he could still imitate as he had with Miles. He succeeded more than 80 percent of the time. But he tended not to imitate tool use in problem-solving situations, as opposed to "Simon Says." Conceding they interpret primate understanding of the intentions of others "more conservatively and skeptically than many other investigators," Tomasello and Call hypothesized that Chantek's failure might have resulted from an inability to understand these sorts of intentions in others. Shumaker disagrees. He thinks orangutans might not be able to imitate a skilled demonstrator. But if they are able to watch a demonstrator learn by making mistakes, they can do it. He also suspects the orangutans in Tomasello and Call's experiment may have been unwittingly taught not to imitate. The reason is they were given the device they used in the experiment to investigate beforehand. Whatever they did with it then went unrewarded.[80]

Anne Russon has videotaped free-ranging rehabilitant orangutans imitating in Borneo. Recently we watched tapes in her laboratory at Glendon College in Toronto starring Supinah, the pancake maker. Supinah enjoyed stealing soap and laundry from the washing staff and trying to do the wash herself. The washers were so terrified of her that they stationed a guard on the dock near the floating raft upon which laundry was done. How to avoid the guard? We watched the video.

A canoe was conveniently moored dockside. Because the orangutans were notorious joyriders, the canoes were usually partly submerged and could be emptied only by sharply rocking them side to side to splash the water over the gunwales. That is what Supinah did. I watched her carefully untie the knot that tethered the canoe to the dock, climb out, and start to rock. At one point, she paused to clamber up the dock and peer over its top to see whether the guard was still at his post. He was, and she returned to the canoe. Finally, she pushed off for the raft and seized the laundry and soap abandoned by the washers, who fled into the river.

Supinah, Russon writes,

> was a prodigious imitator: she hammered nails, sawed wood, sharpened ax blades, chopped wood, dug with shovels, siphoned fuel, swept porches, painted buildings, pumped water, blew blowguns, fixed blowgun darts, lit

> cigarettes, (almost) lit a fire, washed dishes and laundry, baled water from a dugout by rocking it from side to side, put on boots, tried on glasses, combed her hair, wiped her face with Kleenex, carried parasols against the sun, and applied insect repellent to herself. Whenever the job involved a complex technique, hers matched the one used in camp.[81]

The list included imitations of tool use. Russon argues this is more advanced than imitation alone.[82] Two examples: Supinah once tried to build a fire. Happening upon an untended cooking area in which breakfast fires were smoking, she plucked out a burning stick, then meandered to a container of kerosene. After removing its aluminum lid, she took the plastic cup the cooks use for kerosene, scooped a cupful of kerosene, and plunged the burning stick into it. Russon, fumbling and anxious to record the sequence on film, mistook the kerosene for water and remarked that Supinah was trying to douse a fire. A more attentive graduate student suggested that Supinah was actually trying to start one. Imitating the cooks for twenty minutes, she repeatedly tried to light the fire, wet the stick with kerosene, fanned it with the aluminum top, and blew on it.[83] But she never quite succeeded.

Supinah also tried to siphon from a fuel drum into a jerry can. The camp staff siphoned in a uniform and familiar way—unscrew the lids, place the jerry can near the drum, insert one end of a tube into the drum, place the other end into the mouth, suck until fuel flows, then stick the other end into the jerry can. Supinah unscrewed both lids, inserted the end of the tube into the drum, put the other end in her mouth, and either sucked or blew; Russon couldn't be sure. She dropped one end of the tube, then pushed the other end deeper into the fuel drum. Withdrawing the tube, she tasted it, lifted the empty jerry can to her mouth, and then put it down. She stuck the end back into the fuel drum, stepped on it, stuck the other end into her mouth, and sucked (or blew) again.[84] She never siphoned any fuel. No sane human would have allowed Supinah near a drum that actually contained fuel. But Supinah didn't know the drums were empty. Russon points out that this strengthens the case for imitation. As she never actually practiced siphoning, she could only have learned her technique by watching.[85]

Intentional Deception and Pretense

Lyn Miles investigated Chantek's deceptions over seven months of his fifth year and uncovered eighty-seven candidates' acts.[86] She devised a five-level scheme for classifying signing deceptions. At Level 4, we reach intentional deception. The signer intends to manipulate another's behavior. He understands what his signs mean and misuses them to achieve an inappropriate goal because he understands that someone will interpret them correctly. This may require him to suppress cues that give the game away. Level 5 deception occurs when a signer manipulates another's behavior by misusing signs to achieve an inappropriate goal and tosses a red herring onto the linguistic landscape.[87]

Chantek might sign "DIRTY" to visit the bathroom, then not use the toilet. Instead, he would play with the washer, dryer, and soap dish. This is a nightly occurrence with three- and four-year-old Christopher and Siena. After I got them into bed, one calls out every fifteen minutes, "I need to go peeps!" The other chimes in, "I need to go peeps, too." And off they trundle to the bathroom. If I don't watch them closely, they just unroll the toilet paper and scatter the toothbrushes.

Chantek's deceptions weren't all intentional and didn't all reach Level 4. But some did. Twice he signed "MONKEY," so his caretakers would let him into the office in which Miles kept a large toy monkey. But she also stored tools there that Chantek coveted but was not permitted to touch. Once inside, of course, he ignored the monkey and beelined for the tools.[88] Miles says that Chantek showed Level 5 deception, too. He stole a caregiver's eraser. Then he pretended to swallow it, opened his mouth to show that it was gone, and simultaneously signed "FOOD-EAT." In truth, he had secreted it within his cheeks. It turned up in his bedroom where, Miles says, he usually stashed his swag.[89]

Chantek figured out that he could reverse roles in "Simon Says." At Zoo Atlanta, he signed to Miles "SIMON SAYS, GO CAR, GO STORE, GET CHEESE-MEAT-BREAD" (the last, Chantek's description of a cheeseburger).[90] But he usually employed "Simon Says" deceptively to gain an advantage, as when he lured an inexperienced caregiver who carried a whistle he coveted into a faux-game of "Simon Says." Chantek kept signing "DO THE-SAME-THING," followed by an action, until he got around to inducing the caregiver to place his hands on the bars that separated human from orangutan . . . but not by much. Mission accomplished.[91] Sue Taylor Parker says that Chantek's asking his caretaker to "DO THE-SAME-THING" was something that human children don't attain until about age four.[92]

A film crew seeking to capture Chantek playing "Simon Says" with Miles at Zoo Atlanta subjected him to an interminable, tiresome game. Chantek has figured out how to signal when he's through. Some days he signs "FINISHED." That day he reversed roles. Sitting, he twice signed "SIMON SAYS" do this, then do that. Miles obliged. He signed "SIMON SAYS" again. This time, he raised both legs. Miles was standing. She jumped to try to comply and almost fell over. But the message got through. It was, she said, "an orangutan joke, a way for Chantek to say 'enough filming for today.'"[93]

Anne Russon argues simple pretending involves deliberately behaving as if something were real when it's not or behaving as if one were doing one thing when doing something else. More complex pretending requires that one deliberately, and without confusion, hold two different representations in mind, one real, one pretend, and know which is which.[94] Pretending can get complicated. Psychologist professor John B. Watson devised a twelve-step scale to measure pretending in children. Pretending to drink from a cup is an example of Step 1. Pouring water into a doll's mouth is Step 2. Doing the same to a block is Step 3. Having a doll drink from a cup as if the doll were drinking is Step 4 (mental representations are required). Having a block drink from a cup as if the block were drinking is Step 5. Having a doctor doll use a thermometer while a patient doll takes to her sick bed is Step 7. Having a doctor doll examine a patient doll while the child acts as the patient doll's father but also responds to the doctor doll as the patient and to the father as the daughter is Step 12.[95]

At twenty months, Chantek pretended to feed a toy animal. Two months later, he pretended he had to urinate so that he could play in the bathroom. At twenty-five months, he pretended to be afraid of an invisible cat. At forty-five months, he signed "LISTEN" and pointed, in the absence of any strange sound, to avoid going to sleep. At fifty-eight months, he pretend-played with "CLAY MAN," one of many plastic figures, stuffed animals, and other toys with which Chantek played.[96] As human children do, he treated them like dolls, fed them, put them to bed, groomed them, sat them in chairs, ran around in front of them, and pretended they were chasing him. Miles might ask Chantek to "GIVE CLAY MAN DRINK" and he would hold Clay Man's mouth to a cup that sometimes held liquid and sometimes didn't.[97] Chantek's role reversal in "Simon Says" is Step 7 on Watson's scale of child pretense.[98]

Anne Russon doesn't think orangutans who temporarily blind themselves in play are necessarily "pretending," but that the necessary imagination operates at the same level of cognitive complexity.[99] She has written about four cases of orangutan pretense and deception in the wild.[100] One involves Princess's breaking and entering a bunkhouse through an opening that for hours she ignored while pretending to socialize with humans.[101] Two others were orangutan tricks played on other orangutans to obtain food or make contact with them. I will describe her fourth example, because Russon showed it to me on video.

A field assistant, Ucing, was carving a crude walking stick with the large cutting blade of a Swiss Army knife. A feisty, mischievous female named Unyuk, perhaps eighteen years old, looked on in what Russon said was the typically intense concentration of a fascinated orangutan. Her eyes carefully followed the knife's ups and downs from just a few inches away. After a while, Ucing switched to the saw blade and cut off the ends of the stick, then reverted to the knife to scrape off the bark.

Unyuk lifted two sticks and began scraping and sawing one with the other as Ucing had been scraping and sawing. Ucing unfolded the scissors from the knife and pretended to cut Unyuk's hair from the top of her forehead (where she couldn't see what he was doing). Or maybe he snipped a few hairs; it was hard to tell. When Ucing stopped snipping, Unyuk placed the two sticks in a "V," grabbed a hank of hair from the top of her own head, and began to "cut" it with the two sticks.

Ucing returned to carving the walking stick, slicing up and down beneath Unyuk's stare. A moment later, Unyuk's eyes flickered past the knife and her head slightly jerked. She stopped following the knife and focused off-screen. Russon said that Unyuk was peering at the backpacks that she and Ucing carried. They were lying behind Ucing. They usually contained food and are prime targets for orangutan pilfering.

For the next few minutes, Unyuk feigned playing with Ucing, sometimes keeping her head facing the rising and falling knife as though she were watching intently, sometimes holding out her arm or head to encourage Ucing to cut it.[102] Twice she subtly, casually shifted position toward the backpacks. Soon, she was no longer sitting in front of Ucing but was at his side; he no longer sat between her and the backpacks. Then she lunged. Russon jiggled the camera but managed to record the screams attendant to the struggle for the packs.

Once she spotted the packs, Unyuk's goal changed from watching something interesting to larceny. She began to hold two representations in mind at the same time. Posing as the innocent, she pretended to watch Ucing. Because she feigned a complicated role, distracted Ucing by pretending to watch his knife-work, held her arm out for him to cut her hair, gestured, subtly repositioned herself in a way that failed to draw suspicion—all the while behaving like the pickpocket who stole my wallet in a Barcelona subway station—Russon argues that Unyuk's deception involved symbols and qualified as Level 4, the highest level of deception available for nonsigner animals.[103]

Dan Shillito says orangutans at ThinkTank sometimes appear to be trying to thrust their arms through the bars of their cages past the elbow. If a naive human ventures too close, he will learn orangutans have no trouble reaching all the way to the shoulder. When primatologist Karyl Swartz worked with Bonnie at the language board, the orangutan would point to a symbol with one hand and secretly reach under the board to try to get at Swartz with the other.[104]

Orangutan Language

Wild orangutans intentionally communicate with gestures.[105] There have been three projects in which orangutans were taught signs. The first orangutan language project, and the only one in which any nonhuman has been free-ranging, began in 1978 with Rinnie. It ran almost two years. Rinnie was an ex-captive, between ten and twelve years old, being rehabilitated in Borneo. Her home range was just across the river from camp.[106] Carrying the day's materials, Gary Shapiro would swim or paddle across, locate a dry spot, and call Rinnie's name if she wasn't about. Most days she was. She would sit beside or across from him. Lessons might last for a couple of hours.[107]

They were modeled on the techniques for teaching signs to chimpanzees pioneered by Beatrix and Allan Gardner and Roger Fouts. If the name of an object was being taught, say, "banana," Shapiro held up a banana to Rinnie and asked, simultaneously, in American Sign Language and Indonesian, "What is this?" If Rinnie responded with the correct sign, Shapiro gave her food or drink. If she failed, Shapiro molded her hands into the proper shape.[108]

Over twenty-two months, Shapiro taught Rinnie thirty-two words, including "Rinnie," "hug," "flower," and "more." About half he considered to have been formally acquired, because she used them at a high enough frequency. But Shapiro thinks this underestimates her actual competence because she often used more signs when she wasn't being tested.[109] Rinnie also invented signs for "groom," "scratch," and "give."[110] Many signs concerned food. As wild orangutans spend most of their lives searching for and eating food,

Shapiro finds this unsurprising.[111] Russon says "everything reminds orangutans of food."[112]

Chantek's language lessons went on for nearly a decade. Miles's purpose in teaching him "Pidgin Sign English," which uses the signs of American Sign Language in a simplified English word order, was not to have him develop a large vocabulary but to study the underlying mental processes of language acquisition.[113] Chantek learned 150 signs, along with a wide range of linguistic abilities, including a rudimentary grammar.[114] Just as Irene Pepperberg does with Alex, Miles responds to Chantek's signing "as if" it were meaningful.[115] Chantek was encouraged to use signs spontaneously to chat and comment about things and daily events, such as taking a walk or putting a puzzle together.[116] And he did. Eventually, when he wanted to learn an object's name, he asked to be taught the sign by holding out his hands for molding.[117]

Miles's small team began teaching Chantek signs by molding his hands and associating each sign with an object, event, person, or concept, much as Gary Shapiro taught Rinnie signs as they sat on the riverbank.[118] At fifteen months, Chantek began to acquire signs by imitating a caregiver, now without hand molding. His first imitated sign was "WIPER." Made by wiping the hand across the mouth, it referred to a napkin or the need to wipe one's face. At thirty-five months, he showed delayed imitation of signs, first producing them days after he had watched them. At forty months, he imitated a photograph of the signing gorilla Koko pointing to her nose.[119]

When can one be said to have learned a sign? Ape language researchers usually use three criteria. Allan and Beatrix Gardner employed the strictest criterion teaching the chimpanzee Washoe. If Washoe used a sign spontaneously and appropriately on fourteen consecutive days, she "acquired" it. Penny Patterson, who works with Koko, thought the drill necessary to meet the "Gardner criterion" made learning cumbersome and unduly strict. Koko appeared to agree. After drilling her on the signs for the face, one teacher asked what Koko thought was boring. "THINK EYE EAR EYE NOSE

BORING," was her response.[120] Patterson considered a sign acquired if two independent observers noted its spontaneous and appropriate use on half the days of any month. Easier than the "Patterson criterion" is the "emitted criterion"; this includes every use of a sign.[121] Patterson reminds us that once a certain number of words are acquired, an increasing number are naturally used less frequently and it becomes progressively harder to meet either the Patterson or the Gardner criterion.[122] Miles used Patterson's criterion for acquiring a sign.[123]

Chantek rapidly learned signs for objects and actions in which he was keenly interested. Unlike Rinnie's, less than a quarter of Chantek's signs had anything to do with food or drink. But he had the orangutan's obsession with food. He learned "RAISIN" and "BERRY" in just two and a half weeks. Inedible and strange, "HAT" took him over two years.[124] Relational signs—"UP," "DOWN," "IN," and "OUT"—were hard to acquire because they are so abstract. And the same complications were present as they had been for Washoe and Koko. Chantek loved to be picked up and hated to be put down; he learned "UP" in 131 days. But it took 934 days to acquire "DOWN," though it was clear to Miles that he knew exactly what it meant long before.[125]

Chantek's hardest signs were the same words with which Christopher and Siena struggled, pronouns. Miles began teaching "ME" at the age of thirteen months. "ME" is formed by pointing to oneself. "ME," along with "YOU," took Chantek most of two years to acquire. Miles thinks Chantek first used "ME" not to refer to himself but to tell others what he wanted them to do—tickle "ME," chase "ME," give "ME" food.[126] It was a confusing sign to learn because it was the first that referred to different animals, depending upon who used it. If Miles used "ME," it referred to Miles; if Chantek used it, it referred to Chantek. He initially avoided the difficulties inherent in such questions as "WHO GO, YOU OR ME?" by responding, "CHANTEK." But eventually he got it right.[127] Similarly, he used "CHANTEK" for several months before he passed an MSR test. Miles thinks this suggests that he used "CHANTEK" the way he used "ME" when he wanted someone to do something to him, such as chase or tickle, but that "ME" didn't function as a name for him until he had achieved self-recognition.[128]

Chantek began combining signs, such as "MORE FOOD-EAT CRACKER," during his second month's education.[129] By the age of four, nearly 20 percent referred to a thing or event that was not present. More bashful than my twins, who might yell "Food!" and throw open the refrigerator door, Chantek might politely sign "CEREAL FOOD EAT" just before breakfast or while pointing toward the refrigerator. Or he might sign "ICE-CREAM CAR RIDE" as he tugged Miles toward the parking lot, just as Siena drags me to the television set to watch her beloved "Barney" tapes. By the age of eight, these were almost 40 percent of his signs.[130] Chantek also invented signs: "NO-TEETH" (his promise not to use teeth during play), "EYE-DRINK" (a contact lens solution), and "DAVE-MISSING-FINGER" (for Dave who was . . . well).[131]

At ThinkTank, in the third project in which orangutans are taught signs, participation is voluntary. When the orangutans want to learn, they are taught. They have learned about seventy abstract symbols so far, in seven categories—names of orangutans, names of humans, foods, objects, verbs, adjectives, and numbers, as well as simple syntax. The day I visited, Azy was learning 1, 2, and 3.[132]

Azy and Indah, another ThinkTank orangutan, barter. Beck used to give them bamboo pieces to use as tools in public demonstrations. When the demonstration ended, he wanted the bamboo back. Now the orangutans had a bargaining chip. They would demand a sunflower seed for a small piece of bamboo or a walnut for a large piece. When Beck complied, he got the bamboo back. Then the orangutans began to trust Beck. They stopped bargaining. Instead, they handed over the bamboo and, with outstretched hands, expected him to honor the agreement. But bargaining was what interested visitors. Beck could only start the bargaining again by violating the orangutans' trust and breaching the agreement. That he wouldn't do. Those demonstrations stopped.[133]

Theory of Mind

When Chantek or Dona alternates gazes, they are not just intentionally communicating: they must want someone to do something about what they're looking at or at least to share their experience.[134] Like the subordinate chimpanzees in the experiments we discussed in Chapter 1, they understand they can see what others can't.

In *Rattling the Cage*, we saw how Indah's visual attention was tested at ThinkTank.[135] A surprised Dan Shillito watched her reach through cage bars and maneuver Rob Shumaker into a position in which he could see food that he hadn't been able to see before. Indah wanted that food. Intrigued, Shillito and Schumaker constructed a series of experiments to determine whether Indah understood that eyes connect internal states of attention to the world. When confronted with one human whose eyes were covered and another who could see, would Indah manually move the sighted human, but not the blinded one, toward the food? When formally tested, Indah appeared to understand that humans who did not have buckets over their heads could see and therefore help her get food, and would maneuver them into a helpful position. She seemed equally aware that humans who did have buckets over their heads could not see and therefore could not help her, and so she would even move a man who didn't have a bucket over his head around another who did. She might even remove a bucket from someone's head, then prod him into a position to see food, especially if he was closer to the food than a person without a bucket was. And when the researchers sawed off the bottom of a bucket so that when put over the head, it didn't block the eyes, Indah treated the researcher as if he could see.

When he was about four years and ten months, Chantek, before signing, started manually moving his caretaker's face so that his eyes were looking into Chantek's eyes.[136] At four and a half, Chantek began to represent the world from another's perspective. He filched that eraser. By pretending to swallow it and simultaneously signing "FOOD-EAT," he appeared to take his caregiver's visual perspective and understand that if he hid the eraser within his cheeks, the caregiver would not be able to see it. This was central to his deception of the caretaker into believing that he had swallowed the eraser.[137] When Chantek finally learned that "ME" referred to himself when he used it, and to others when they used it, he was able to represent the world from the perspective of those others.[138]

Symphonic Events

Some of Russon's videos show free-ranging rehabilitant orangutans uniting multiple abilities; this she calls "one symphonic event."[139] We saw that Unyuk simultaneously demonstrated at least four complex mental abilities: deception, imitation, complex gestural communications, and tool use. That day in Toronto we viewed a third symphonic video.

Nine-year-old Siti tried to induce a forest assistant, Nian, to open a coconut. Bornean coconuts are unlike coconuts in my supermarket, which have a thick three-eyed shell and a large central chamber that holds milk and meat. The only edible part of a Bornean coconut is the tiny jelly contained in the chambers to which the three eyes lead. These are hard to enter.

Siti chewed off the thick fibrous husk that concealed the eyes, then poked a finger into the chamber of one eye. Picking up a branch, she stripped a slender length from which she fashioned a stick to probe the chamber more deeply through the opened eye. Only partly successful, and leaving two eyes untouched, she handed the coconut to Nian. Here Russon chuckled.

"It's against the rules," she said. "Staff are not allowed to open coconuts for orangutans."

But someone had aided Siti before. Before Russon's softly whirring camera, Nian returned the coconut and sternly advised Siti to open it herself. He gestured toward a stick. She used it to poke for a few moments, then handed the coconut back. In case Nian didn't get the picture, she drew one by making vigorous hacking gestures across the coconut with the stick. Russon found Nian's discomfort and Siti's insistence vastly amusing.

In just five minutes, Russon said, Siti had displayed numerous cognitive abilities. She used a complex food-processing technique to get inside the coconut, then made a tool to help out. By hitting the coconut with the stick, she improvised iconic gestures that unmistakably communicated to Nian what she wanted him to do. Here she displayed an understanding of cause and effect. She demonstrated logical equivalence by hacking the coconut with a substitute machete—the stick—to show what she wanted. She apparently imitated what she had seen him or someone else do with a machete. Russon thought it no more likely that anyone would teach Siti to use a machete than allow Supinah to siphon fuel from a jerry can. Siti tried to use Nian as a social tool. Finally, by modifying her message when Nian didn't comply with her request, Siti exhibited "intersubjectivity," the interplay between conscious minds.[140]

Does Chantek Have Practical Autonomy?

Chantek is a Category One animal, in practical autonomy not lacking much that a chimpanzee, bonobo, or young human has. I assign him an autonomy rating of 0.93. Very few orangutans live in captivity, and together they have negligible economic value. They are indigenous only to Africa. An evolutionary hop, step, and jump from human beings, compared to Phoenix and Ake, Chantek's brain resembles ours in vital ways that the dolphin brain does not. Chantek also possesses a powerful self-awareness, having first passed an MSR test at age twenty-five months, and reliably passing them by age three and a half. His MSR is more established than either Echo's or Phoenix's and Ake's. He inspects his face and makes faces in mirrors, uses mirrors to help him curl his eyelashes, and names his mirror image while conversing with others or signing to himself. He possesses both James's

"Me-self" and "I-self," Damasio's autobiographical self, and not just Neisser's ecological and interpersonal selves but a conceptual self, perhaps a temporally extended self, possibly even a private self.

From his abilities to respond to Miles's request that he "SIGN BETTER," to use tools in complex ways, to manufacture tools and tool kits, to use tools insightfully so that he can mentally see a problem's answer without engaging in extensive trial and error, and, with foresight, to plan goals and routes, and to pass Piagetian Substage 6 object permanence tests at the level of a human two-year-old, it is clear that he mentally represents. He is as competent in the conservation of liquid quantity as human seven-year-olds.

He intentionally communicates by alternating his gaze from the eyes of the human with whom he is communicating to a place he wants that person to look; and by pointing, he understands that humans are intentional agents. He imitates precisely, and this has been critical to his ability to learn sign language as well as he has. He intentionally deceives and engages in complex pretense, including pretend playing and role reversal, which Phoenix and Ake have not demonstrated. He learned a simple sign language of 150 signs marked by a rudimentary grammar and the use of "ME," "YOU," and "CHANTEK." He uses his signs spontaneously and to comment. He often refers to objects not present and appears able to see at least some of the world from the perspective of others, if not to the degree that chimpanzees and bonobos may, again an ability Phoenix and Ake have not shown.

ELEVEN

Koko

"Mike, Mike"

For twenty years I tried to visit Koko and Michael, the world's only signing gorillas, at their home in Woodside, California. But Francine (Penny) Patterson, who began Project Koko almost thirty years ago, would be leaving town as I arrived. Or Koko would have a cold. Or my trip was canceled. Nobody's fault. Each time, I happily submitted to a tuberculosis test and learned I did not have the TB to which gorillas are highly susceptible. Then, on April 21, 2000, I opened the *New York Times* and discovered that I would never visit them both. Michael had died two days before from a heart attack, just a month after my last attempt to visit. It was no great surprise. More than 40 percent of captive gorillas die from cardiovascular disease.[1]

Orphaned at age three in Cameroon when his parents were killed and eaten, Mike had been in his silverback prime. The *Times* reported his death so: "One of just two gorillas in the world who are said by handlers to have learned human language has died. The lowland gorilla, called Michael, was 27 and, his handlers said, knew about 500 signs and gestures in American Sign Language, which is used by the deaf in North America. . . . Experts are divided over whether apes can use language; critics maintain apes who use gestures or symbols are merely imitating researchers to get rewards or are responding to humans' unconscious cues."[2] The day after Mike died, Patterson told Koko what had happened. Koko responded with both hands simultaneously.

"MIKE, MIKE," she signed.[3]

He had been her intended mate, but it had just never worked out. Instead, he became her playmate, conversation partner, occasional pupil, boon companion, and, alas, baby brother. She may never get over his death. In 1984, her kitten, All Ball, was struck and killed by a speeding car. Four months later, Patterson asked her, "How did you feel when you lost Ball?"

"WANT," Koko said.

"How did you feel when you lost him?"

"OPEN TROUBLE VISIT SORRY," Koko said.

"When he died, remember when Ball died, how did you feel?"

"RED RED RED BAD SORRY KOKO-LOVE GOOD."

As is every species of great ape, gorillas are endangered, some subspecies critically so. There are perhaps 100,000 individuals remaining in the wilds of central Africa. As are chimpanzees and bonobos, gorillas are being increasingly hunted for their meat. Unless extraordinary measures are put in place, they will soon become extinct in the wild.

"SIT KOKO—LOVE YOU"

I tried again to see Koko. Seven months later, I found myself just a few miles away. She was living just up a wooded path. Penny Patterson had said she would meet me outside the local market at 1:00 P.M. She arrived in a white Ford Explorer with blue waves painted along the sides. The driver of a trailing car waved. I waved back and pulled into line. We convoyed for fifteen minutes along a narrow, steeply rising road, filled with hairpin turns and flooded with pine-filtered sunlight, then turned left and continued rising through the Santa Cruz Mountains until we reached a dirt lane that led into a wood. I parked my car and was introduced to DeeAnn Draper, one of Patterson's employees. I had been told to wear sturdy shoes and avoid plaid.

Koko's other visitor was Al Hillix. He was coauthoring a book on nonhuman language with Duane Rumbaugh, a pioneer in that field.

Absent from home a week, Patterson needed to speak to Koko alone. "You know how Koko gets when Penny's gone," DeeAnn said. I didn't. But Penny nodded grimly and headed up the path.

We climbed past a mound of pottery shards. Ann said the woman who once owned the property had buried pottery everywhere. That didn't seem as odd then as it does now. Sturdily shod, clothed in blue jeans and blue sweatshirt, I was climbing to chat with a gorilla who knew some American Sign Language (ASL) and understood some English. She would understand some of what I said. I would need a translator. In a hundred yards, we entered a low building and were introduced to another of Patterson's employees, Dina Pettit. Then Penny reappeared with Ron Cohn, cofounder of Project Koko and the man who has carefully documented every milestone on film. All was now well with Koko. She knew we were coming. I learned we were Koko's first visitors in three months; the last had been Robin Williams and Sting.[4] Penny led us to a small porch that jutted from the center of a shaded yellow trailer trimmed with brown wood, Koko's bedroom for almost thirty years. Dina hung back, took out a pen, and began to write. I took out my pen.

Koko has appeared in numerous documentaries, books, and magazines, including *Playboy,* where she responded to "20 Questions."[5] I recognized her instantly. Five feet and three hundred pounds, she sat pot-bellied against the mesh that separates her from her visitors.

"COME-ON DO YOU," Koko signed. "LOWER FAKE TOOTH" to me and "UPPER FAKE TOOTH" to Al.

Al had a prominent gap between his upper front teeth. I made a mental note to examine my lowers, which I had thought were nondescript. Koko asked me to open my mouth, wide. Penny said she wanted to see whether I had any gold teeth. She had demanded the same of the *Playboy* interviewer fourteen years before. I opened wide and pointed to the gold crown in my upper left quadrant. Koko purred.

"SIT KOKO-LOVE YOU," she signed.

I sat on the ancient three-legged stool that occupied most of the tiny porch.

"YOU SIT HURRY," Koko said to Al.

I had the stool, so Al had to kneel. Penny handed Koko a treat.

"CANDY YOU," she said to Al. "UPPER FAKE TOOTH."

He had no candy.

"CHASE," she said. I knew that meant she wanted me to chase her by running after her as she raced the length of her trailer and back.

Koko moved to the toilet in the back of the trailer, plopped onto its red seat, and loudly urinated.

"Visitors make her nervous," Penny said.

I began to ask Penny a question about Koko.

"You really should speak to her." Penny nodded toward Koko.

I waited until she returned from the toilet. There followed an hour of chatting, chasing up and down the trailer's length ("TIME YOU," "CHASE DO YOU," "YOU CHASE," "DO CHASE HURRY DO"), tugging on blankets, requests to enter the kitchen portion of her trailer ("UPPER FAKE TOOTH THERE"). Ron Cohn produced a cup of coffee.

"DO YOU," said Koko.

Penny handed her the cup. Koko removed the plastic cap and began to drink.

"DO CANDY," she said to Penny.

Penny disappeared and returned with a bowlful of sliced cooked vegetables, one of Koko's seven daily meals. Everything swam in pink.

"What is this stuff?" she asked, showing the contents to Koko.

"DO THAT," Koko said. "MEAT THERE."

Koko gestured toward the bowl. Penny said to me she thought it was only vegetables—no meat—despite the pinkish glow, but that it smelled of ginger. She handed small pieces to Koko through the wide mesh.

"SQUASH," Koko said. "POTATO. ONION. BROWSE (for broccoli). BERRY."

Penny said she thought it wasn't a berry, but a beet. She held up a sliver of garlic. "What's the sign for garlic?"

"BERRY DO," said Koko.

Penny turned to me.

"'BERRY' might have actually been 'BERRY,'" she said. "But it could have been 'BEAN.' The signs look somewhat alike."

She handed Koko a drink. When Koko tugged at something in her teeth, Penny opened her purse and removed a small thin mirror, then held it so Koko could see what was caught. Koko leaned forward and peered at the mirror into her mouth, plucking at a piece of vegetable. Then she handed Penny a speck of ginger. After finishing her snack, she chose an alligator puppet from the dozens of plastic alligators, lizards, lions, dolphins, sheets, blankets, and other objects that littered her bedroom floor, then ambled toward the kitchen. Penny invited me to "get scared." I walked over. Koko brandished the alligator through the glass panel that separated the kitchen from Koko's living quarters.

"An alligator!" Penny said in mock fear. "I'm afraid!"

Then Koko "scared" me.

"An alligator!" I said, trying my darnedest to look terrified.

"DO THERE COME," Koko said. "CHASE DO YOU."

Then she wanted me out of her kitchen.

"THERE THERE HURRY."

I rejoined the group huddled on the tiny porch as Koko raced past, waving the alligator. She spied my cell phone.

"LISTEN," she said, then "VISIT DO YOU" to Al.

She noticed a pair of eyeglasses he carried.

"BRING THERE."

I asked Penny how rapidly Koko could sign. Slowly with novices, she replied, much faster with herself. We prepared to leave.

"NIPPLE," Koko said.

Penny explained that Koko enjoys staring at her visitors' nipples. I have no idea what Sting did, or Robin Williams. But Al and I obediently displayed our nipples, he by unbuttoning his shirt, I by raising my blue sweatshirt.

"Good-bye, Koko," I said.

"NIPPLE," Koko replied.

"We're finished with that," Penny scolded.

Koko rumbled back to the toilet. We stayed until she had finished.

"DRAPES," she said.

As Dina closed the drapes of the trailer, Koko blew us a kiss.

Don't be fooled by Koko's terseness. In their path-breaking study of American Sign Language, linguists Ursula Bellugi and Edward Klima found that humans sign telegraphically, too. The sentence "It is against the law to drive on the left side of the road" might be signed "ILLEGAL DRIVE LEFT-SIDE." Fourteen words condensed to three signs. Consider, they said, the following story: "They both looked at me. And they looked at each other. And then they started laughing and laughing. This made me burst out crying." This might be conveyed as: "TWO-OF-THEM LOOK-AT. THEN LOOK-AT. START LAUGH. MAKE (ME) CRY."[6] Nine signs, twenty-three words.

Koko and I last shared an ancestor about 8 million years ago.[7] Hold your breath for less than three seconds and you understand that she is a close evolutionary sister, nearly as close as the bonobo and chimpanzee. That is why there is little doubt that Koko is not just intelligent, but intelligent in a human way. Whatever intelligence tests measure, Koko has at least some of it. It's not straightforward to compare the intelligence of humans and nonhumans based on their scores on tests designed for human children. Nor is it even clear that tests designed for human children are valid for nonhumans.[8] But on a battery of ten standardized tests for human children administered over years, Koko consistently registered an IQ of between 70 and 95. This placed her in the "slow," but not retarded, range; her mental age increased from 10.8 months (when Koko was fourteen months old) to 4.8 years (when she was five and a half years old).[9]

Gorilla Sign Language

At the Madrid Zoo, psychologist Juan Carlos Gómez, from the Universidad Autónoma watched a group of five gorillas grow from infancy. He confronted them with interesting situations. Would they would use tools, inanimate or social, to try to solve the problems intentionally?[10] If they persistently tried to surmount an obstacle to reach a goal, then stopped trying when they succeeded, he counted them as having done just that.[11]

Gómez hung a swing coveted by one young female, Muni, too high for her to reach. She tried various strategies. All failed. When a human stationed himself beside the swing, Muni approached him, lifted her arms in his direction, looked into his eyes, and froze her posture until he lifted her toward the swing. Muni then turned and reached for it.[12] This was communication used to solve a problem, and it began at about the age of one year.[13]

Another time, Muni moved a box under a door latch, stepped onto the box, and began to unlatch the door. Gómez stopped her and removed the box. Soon, Muni approached, took his hand, and pulled him back toward the door. As they walked, she alternately looked at the door and into Gómez's eyes. At the latch, she gestured to be picked up. When Gómez complied, she reached for the latch.[14] Muni often approached Gómez for assistance, touching his knee until he looked down and made eye contact, grasping his hand, then leading him to where she wanted him to be, usually engaging in eye contact as they moved together.[15] The gorillas might not just touch his knee to get Gómez's attention, but pull his clothes, tap his hand, physically turn his head toward their eyes, throw straw at him, even grunt. Whenever Gómez responded by looking into a gorilla's face, she stared back.[16] Muni often distinguished between requests that were received and declined ("No!"), which she tended not to repeat, and those her intended recipient didn't seem to notice.[17] Gómez thinks she understood that his attention and actions were linked. She could implicitly "read" his eyes and gaze direction, his facial expressions, and body postures. This allowed her to communicate intentionally, even if she lacked the ability mentally to "read" his mind.[18]

Primatologist Joanne Tanner and her husband, Charles Ernest, noticed something more while observing half a dozen gorillas at the San Francisco Zoo.[19] They observed the gorillas using natural "iconic gestures," head or limb movements that follow the path of an action that one desires to communicate when socially interacting.[20] A condemned prisoner who watches his sovereign trace an index finger across his throat is seeing an iconic gesture. Nothing so dramatic went on at the San Francisco Zoo. When Mkubwa (Koko's brother, known as Kubie) wanted to be touched, he might gently and quietly pat his chest with a partly closed hand as a young female, Zura, watched. This gesture often induced her to touch him during play. If Kubie swept a hand along Zura's back, she often moved in that direction.[21] Evidence shows that wild gorillas communicate direction in the same way.[22] As we saw earlier, Koko has moved far beyond iconic gestures, although her signing may be rooted in them.[23]

Comparing the vocabularies of apes and human children often leads to an underestimation of the apes, because they are usually compared to hearing, speaking, human children. In a comparison with a hearing child, Koko's language at forty-one months resembled the child's at twenty-four months.[24] Their differences might be attributable to species. But they might also stem from variations between spoken and signed languages.[25] To test this, Patterson compared the sign acquisitions of Koko and Michael to twenty-two human children, most under the age of one. Two were deaf. Seventeen had two deaf parents; the rest had one.

In many ways, Koko's early sign acquisition, from the one-word stage on, resembled that of young, especially deaf, children.[26] Each used the same types of signs: proper nouns (PENNY, KOKO), pronouns (ME, YOU), nouns that stood for living creatures (BIRD, DOG), nouns for food (BERRY, POTATO), other nouns (BALL, LIPSTICK), adjectives (ORANGE, SMALL), negatives (CAN'T), verbs (EAT, TICKLE), and prepositions (ON, UP). But Koko did not consistently use interrogative signs, the "WH"s (WHO, WHAT, WHERE), as human children did, though she did on occasion. Almost from the beginning of her education, however, she questioned by holding a sign steady, cocking her head, and seeking direct eye contact.[27] She regularly responded, correctly and appropriately, to "WH" questions. When asked "WHO YOU?" for example, she responded "KOKO." To "WHAT COLOR THIS?" she said "THAT RED."[28]

The chief differences between the gorillas and the children in Patterson's study were their rates of acquisition; children acquired signs faster than gorillas.[29] That could be for many reasons. There were obvious rearing differences. The human children were not born in the wild, their parents were not murdered before their eyes, and they did not suffer kidnappings that took them to three continents, as happened to Michael. The human children's primary caregivers were their parents, who were almost always fluent signers. Koko and Michael were eventually raised by Penny and a large number of assistants, many of whom were not proficient in ASL. The human children also received sign language instruction very early on. Koko didn't see her first sign until she was a year old; Michael was three and a half.[30]

Still, Koko signed her first word within days of beginning her schooling.[31] When Koko was forty-one months old, Patterson filmed a one-hour dinner conversation during which Koko used 251 signs.[32] By 1982, she had acquired about 290 signs according to the Patterson criterion, 200 by Gardner's; she had emitted 876 signs and actively used about 500.[33] Just 250 signs may be enough for a deaf human child to start school, and the core vocabulary of a fluent signer may total from just 500 to 1,000 signs. That's because a small number of signs can be modulated to express many more ideas. Using modulation, signers add meaning by changing the location, shape, or motion of a sign, or of their own body posture or facial expressions while forming a sign.[34]

Koko uses at least ten modulations.[35] She might change the size of her signing space. ALLIGATOR is made with two hands snapping shut. She may snap her arms shut to refer to a large alligator. For a small alligator, she might snap just her fingers.[36] She has signaled how very thirsty she is by signing THIRSTY from head to stomach instead of just down her throat.[37] By moving SIP toward another's lips, she says "YOU SIP."[38] She began to use NUT to mean someone acting in a silly way by placing the thumb from a fisted hand to the side of her mouth, instead of the middle, where NUT (the food) is usually placed.[39]

Koko may combine two signs, as when she said "SIT KOKO-LOVE YOU" to me. She may combine three, even four signs, simultaneously, such as HURRY-POUR-THERE-DRINK.[40] She does this three ways, using two hands (remember "MIKE, MIKE"), adding one sign's motion or shape to the place in which another sign is being made, or adding a facial expression or motion of her head to a sign.[41]

No one taught Koko English. Yet by 1993, she could understand several thousand English words. Test results suggested she understood English at least as well as sign language.[42] Koko also responds to written English. She has begun to read simple words and a few numbers. Perhaps this should not be surprising; the hundreds of ASL signs that she has learned are visual symbols not that different from the printed symbols she is learning to read.

Koko has progressed through at least four reading-readiness stages. She began by convincing teachers that she understood what written words symbolize. Then she asked them to write words in which she was interested. She seemed to understand some two- and three-word sentences. Finally, she realized that words are constructed from letters, which, in turn, have sounds that can be combined. She developed a spelling system similar to the one I used when piloting small planes. I might partly acknowledge an instruction from an air traffic controller by saying "Bravo Lima." This told them I was flying an aircraft whose identification number contained the letters "BL." Koko uses signs that begin with the letters of words she is trying to learn. She might spell "cat," for example, "CAT APPLE TEETH."[43]

When she was four, Koko took the standard "Assessment of Children's Language Comprehension" test in spoken English, sign, and simultaneously in sign and English. The test asked whether she understood a request such as "point to the bird above the house" and whether she could choose the scene that correctly mirrored a sentence. She performed less well than would a normal human child, at about the level of an educationally handicapped human child of four or five years, but equally well in sign and spoken English.[44]

Koko invented signs for BITE, TICKLE, and STETHOSCOPE. BITE is usually made by placing one clawed hand atop the back of the other. Koko places her mouth on a hand. As there is no ASL sign for

TICKLE, Patterson borrowed the Gardners's, drawing an index finger across the back of a hand. Koko's is more to the point: she draws her index finger across her underarm.[45] She made up a sign for STETHOSCOPE by sticking her index fingers into both ears.[46] She may make action signs—TICKLE or PINCH—on her thigh or stomach to show where she wants the action.[47]

Some of Koko's invented signs are iconic. They so clearly resemble what they refer to that someone who doesn't know sign language would probably understand what they mean. Perhaps 10 percent to 15 percent of ASL signs are iconic. EAT is made by moving a hand to and from the mouth.[48] A bit more than half of Koko's and Michael's first ten signs were iconic, slightly less for their first fifty. As they matured, the percentage of iconic, signs steadily dropped, that of non-iconic signs, which don't physically resemble what they symbolize, grew.[49] Interestingly, Koko has invented iconic signs for English sounds, KNEE for NEED, BROWS for BROWSE.[50]

Her signing follows simple rules, such as placing MORE at the beginning of a sentence or verbs either directly before or after pronouns.[51] Word order isn't random.[52] She says "PINCH SHOULDER," rarely "SHOULDER PINCH." Though her teachers always say "EAT BANANA," Koko says "BANANA EAT," which suggests that she sometimes follows her teacher's instruction, sometimes not.[53]

Koko may sign BITE or TEETH to herself before chomping onto window frames or pens, or THAT TOOTHBRUSH or THAT NUT when she spots an appropriate photograph while leafing all alone through a magazine.[54] Young children often speak or sign to themselves.[55] I installed a monitor in Siena and Christopher's bedroom when they were born. During the third year, Siena often chatted to herself in bed, sometimes for forty-five minutes; during dinner, she provided her parents and big sister a chronology of her day's activities.

Koko refers to the past, even to past emotional states. Young Koko once bit a caretaker. The following day brought this exchange:

"What did you do yesterday?" Patterson asked.

"WRONG WRONG," said Koko.

"What wrong?"

"BITE."[56]

Once Koko bit Patterson. This conversation ensued three days later:

"What did you do to Penny?"

"BITE."

"You admit it?" Patterson asked.

"SORRY BITE SCRATCH."

Patterson exhibited her scratched hand to Koko.

"WRONG BITE," said Koko.

"WHY BITE?"

"BECAUSE MAD."

"WHY MAD?" asked Patterson.

"DON'T KNOW."[57]

Koko rhymes in English. Patterson asked:

"Can you do a rhyme?"

Koko responded, "HAIR BEAR."

Patterson said, "Another rhyme?"

Koko's response? "ALL BALL."

Another teacher asked:

"Which animal rhymes with hat?"

"CAT," said Koko.

"Which rhymes with big?"

"PIG THERE." Koko pointed to a pig.

"Which rhymes with hair?"

"THAT." Koko pointed to the bear.

"What is that?"

"PIG CAT."

"Oh. Come on."

"BEAR HAIR."

"Good hair. Which rhymes with goose?"

"THINK THAT (points to the moose).[58]

In what Patterson suspects might be an early use of metaphor, and is powerful evidence that Koko can mentally represent, she combines signs to describe new objects, SCRATCH COMB for hair brush, COOKIE

ROCK for a stale roll, EYE HAT for a mask, WHITE TIGER for a toy zebra, and BOTTLE MATCH for a cigarette lighter.[59] So did Michael: ORANGE FLOWER SAUCE for nectarine yogurt; BEAN BALL for peas.[60] Patterson thinks that Koko's reference to her feelings, describing herself as a "RED MAD GORILLA" or "RED ROTTEN MAD," provides even stronger evidence.[61]

Koko may use language in insulting and possibly humorous ways. After Michael bit one leg off a Raggedy Ann doll, Koko tore off the other. Patterson scolded Koko for ripping off both. The videotaped rejoinder of a half-wrongly accused gorilla clearly shows Koko's reaction: "YOU DIRTY BAD TOILET."[62]

Here are two samples of Kokoian humor. One of the first three signs Koko learned was DRINK.[63] It's made by placing a fist in the "thumbs-up" position and touching the thumb to the mouth. Koko has appropriately signed DRINK thousands of times. Yet a teacher who had known Koko for years once repeatedly asked her to sign DRINK. She responded with "SIP," "THIRSTY SIP," and "APPLE SIP."

"Koko, please," the teacher implored. "Please sign DRINK for me."

Grinning, Koko leaned against a counter, placed her fist in the "thumbs-up" position, and stuck the thumb . . . in her ear! Here is what supports Patterson's claim that Koko was being humorous: the grin, Koko's legendary stubbornness, and her thorough understanding and extensive prior appropriate use of DRINK. Neither before nor since has she signed DRINK to her ear. And everyone to whom I have told the story, primatologists included, laughs.[64]

Here's the second. While testing four-year-old Koko's imitative abilities, anthropologist Suzanne Chevalier-Skolnikoff asked Patterson to point to her eyes, nose, mouth, ears, forehead, and chin, and request that Koko do the same. Koko complied perfectly. When Chevalier-Skolnikoff sought to film a repeat performance a few weeks later, Patterson pointed to her eye. Koko stuck her face close to Patterson's and seemed to study the request. She pointed to her ear. Patterson pointed to her nose. Koko leaned into Patterson again, contemplated the invitation, and pointed to her chin. This went on for five minutes, when Patterson had had enough. She chastised Koko and signed "BAD GORILLA." Koko's response? A laugh and "FUNNY GORILLA."[65]

There is no penetrating to the center of the thicket of gorillas' intentions. Or of children's. Or husbands' or wives'. No scientist is going to *prove* that anything any human ever said or did is humor. Divining gorilla humor is no easier. The evidence of Koko's humor, like all evidence of human humor, is anecdotal. The idea that Koko might use humor has been criticized: if a nonhuman animal uses language nonliterally, she must simply be making a mistake.[66] But the essence of much humor, as with metaphor, is the departure from the literal, or at least an exaggeration of it. Does Moe really slap Larry and Curly that hard? How can Wile E. Coyote survive repeated falls from such high cliffs? What about, "Take my wife . . . please"? At thirty-seven months, Siena began adding "bump, bump, bump, bump" to each line of a song I have sung to her for years just before she goes to sleep. She giggles and exclaims, "Siena make a joke!" Try to prove she's being humorous. Or isn't.

Tool Use

Wild gorillas have the acquaintance with tools that I have with Jennifer Lopez.[67] Slender data on tool use in captive gorillas once painted all gorillas as rare, awkward, and unhandy users.[68] In the 1990s, observations of spontaneous tool use and tool making by captive gorillas mounted. Scientists observing eleven gorillas for three years at a zoo in Gabon reported the regular use of sticks and plant stems as reachers, sticks as weapons, plants as sponges, and logs as ladders.[69] Elsewhere, a zoo gorilla dragged toward him an out-of-reach object that rested on a cloth that he could reach. But he didn't drag the cloth when that object was placed on the ground beside it.[70] In a third study, researchers placed tree branches inside an enclosure housing fifteen gorillas. Outside, they placed honey and peanut butter in a tray the gorillas couldn't see. The gorillas used, and sometimes modified, the branches to dip into the food; they succeeded in removing the peanut butter and honey in fifteen ways.[71]

In 1994, investigators inquired into the tool use of gorillas held in zoos in the United States and Canada. In deciding what qualified as tool use, they used anthropologist Ben Beck's definition of a tool user as one who deploys an unattached environmental object to alter more efficiently the form, position, or condition of another object, another organism, or the user herself when she holds or carries the tool during or just prior to use and is responsible for the proper and effective orientation of the tool.[72]

Twelve zoos reported that fifty-two of fifty-six gorillas used tools.[73] The scientists asked whether the gorillas used these tools in ten possible ways, including throwing, sponging, probing with a stick, using a ladder or stool, hammering, and raking. Between 15 percent and 60 percent of the gorillas used all ten; about half threw, sponged, or probed. Fifty-two percent modified their tools for use. The zoos volunteered that gorillas used four other kinds of tool, including a stick as a fork and a cup to drink. Fifty-four percent of the gorillas manipulated objects in such complex ways as stacking or placing one inside another.[74] The scientists concluded that tool use and complex object manipulations were common among zoo gorillas.[75]

Gómez saw similar actions at the Madrid Zoo. He confronted ten-month-old Muni with problems she could solve only by moving away from her goal. Gómez would place a desired object atop a long high concrete platform. Muni would move from the object toward a tree connected to the platform, climb it, and walk onto the platform.[76] Within a couple of months, she was using crates, brooms, and poles to solve the problems more directly. By the time she was three and a half, she regularly obtained water from a dispenser that required her to press with one hand while holding a container in the other. She threw straw bundles over objects beyond her cage and dragged them in.[77] Hers seemed intelligent actions.[78]

When she tried to enter a room to which Gómez denied her access, Muni would find the door locked. She might drag her toy tricycle to a point just opposite that pesky latch. Even after mounting the tricycle, the latch would still be beyond her reach. She would dismount, push the tricycle directly under the latch, look between the trike and the latch, remount, and grab the latch.[79]

As they matured, all the Madrid gorillas attained all three circular reactions, primary, secondary, and tertiary, characteristic of Piagetian Substages 2, 3, and 5.[80] Koko achieved them by age four.[81] Gómez conclude that gorillas "are not only sophisticated problem solvers," but "engage in complex explorations of objects."[82]

Early on, Muni used humans as objects in the same way she used her tricycle.[83] If Gómez knelt before the locked door, Muni would grab him by the neck and pull him to all fours. Without looking at his eyes, she would clamber onto his back, stand, and reach the latch.[84] Using Gómez as she did her tricycle was different from communicating with him as an agent able to perceive and act. As we saw, that came later. When it did, Muni emerged from a noncommunicative mechanical world of acts into a world of communicative gestures.[85]

Gorilla Self-Awareness

The video "Conversations with Koko" dramatically begins with a scene in which Koko peers into a large mirror.

"Who is that?" Patterson asks.

"THINK ME THERE," Koko says.

"Okay, that is you," says Patterson.

"GORILLA ANIMAL KOKO LOVE."[86]

When asked "Who are you?" ten times, all ten of Koko's responses contained the words "KOKO," "ME," and "GORILLA."[87] Asked to "Show me your tongue," Koko sticks out her tongue. When she reflects on her emotional states, she says she is THIRSTY, HUNGRY, SAD, or FINE.[88] When she signs BITE or TEETH, THAT TOOTHBRUSH, or THAT NUT to herself when no one else is around, to whom is she signing?[89] When Koko appropriately and consistently refers to herself as "I" and "me" and "Koko," is this not strong evidence that she is self-aware?[90]

For reasons suspected to be as diverse as a lack of interest in mirrors, a lack of experience with mirrors, low motivation, or an aversion to staring, many gorillas flunk the standard MSR test.[91] But not all. Koko and Michael passed.[92] Koko grew up around mirrors. When she was three and a half, she consistently groomed her face and underarms, picked her teeth, made faces, watched herself eat and sign, inspected her teeth and tongue, brushed and flossed her teeth, signed to herself, and put on hats and wigs and makeup while looking into the mirror. Told she had something on her face, she would go to a mirror, locate the blemish, and remove it.[93] She just wasn't administered the standard MSR test until she was nineteen.[94] The ease with which Koko used a mirror to examine her teeth during my visit shows that she is obviously an experienced mirror user.[95]

Unenculturated gorillas have succeeded at informal MSR tests. Scientists showed videotapes to four western lowland gorillas, Delilah and Jason, both twenty-six years old, Diana, age seventeen, and Jeremiah, age five, at the Bristol Zoo in England. Some of the tapes showed unfamiliar gorillas, some were films of themselves taken the day before, and some were video of themselves in real time. The gorillas sometimes reacted to the third category of tapes as if they knew they were seeing themselves in real time. From behind, Jeremiah pushed a barrel toward Diana, who was gazing at her live image. She sidestepped it without a backward glance. Jeremiah examined the inside of his mouth on the monitor. While staring at the live video, Delilah felt along her brow ridge, while Diana touched her face, scratched her nose, and felt along her brow ridge.[96]

Other unenculturated gorillas have passed formal MSR tests. At the age of twenty-two, King, once forced to perform in a circus and onstage, then a resident of Florida's Monkey Jungle, passed.[97] So did Pogo at the San Francisco Zoo. Sue Taylor Parker had set up a mirror there to study MSR. After Pogo had accidentally daubed her face with finger paint, she looked into the mirror, then wiped off the paint with her hand. Parker marked the face of Kubie's father, Bwana, with paint, then saw him swab it with a piece of paper as he stared into the mirror.[98]

Gorilla Imitation

The English language has twined aping and imitating for centuries. *The Oxford English Dictionary* says that "to ape" means "to imitate" or "mimic," and to "imitate, *esp.* in an inferior or spurious manner, to counterfeit, to mimic the reality." In 1685, art was said to "ape nature." "To ape it," as in "[t]he devil loves to ape it after God" (1868), means "to play the ape, mimic the reality." One may play the ape as "an imitator, a mimic contemptuously or derisively."[99]

Richard Byrne studies imitation. Some of the best evidence of imitation in the wild comes from mountain gorillas.[100] They are fond of eating the galium plant, which is blanketed with sharp little hooks; eating one without removing the hooks would be like swallowing a rose bush. Imagine an infant perched atop his mother's head watching her remove those hooks. Now he wants to do it. He might copy his mother's movements with great precision. If he does, this is action-level imitation. Or he might puzzle out how to process the galium himself. That would be emulation. Or he might copy the gist of what his mother is doing and thereby engage in program-level imitation, a more general copying than action-level imitation, but a truer copying of the purpose of the behavior than emulation.[101] Byrne argues that this is exactly what gorillas do.[102]

Gorillas seem capable of action-level imitation, too. In one study, scientists presented eight zoo gorillas with an "artificial fruit" invented to resemble a food-processing task similar to what they might encounter in the wild. This task allowed scientists to study imitation more methodically. A hinged lid and a handle were fastened with bolts or a pin that the gorillas had to remove to reach a banana contained within. Before the gorillas got their chance to try, humans showed them five times how to "process" this artificial fruit by demonstrating different removal sequences. Then these eight, plus two gorillas who had not seen the demonstrations, were given the artificial fruit. The naive duo never succeeded in opening the artificial fruit. Two of the educated eight didn't care to play the game. But the other six did succeed, and in between 12 and 298 seconds. Because only those gorillas who had watched the demonstrations opened the artificial fruit in just the way it had been demonstrated, the scientists concluded that they were engaging, at least in part, in action-level imitation.[103]

From early on, Koko imitated like mad.[104] It would seem impossible to learn 1,000 signs any other way. She could learn signs for things she liked—BERRY, SWING, SOAP, PINCH—in a few minutes. Disliking eggs and hand lotion, it took her months to acquire EGGS and LOTION.[105] By the second year, Koko was imitating many of Patterson's actions, such as speaking on the telephone and manicuring her nails.[106] When a policeman mimicked the sound of a galloping horse, Koko imitated his clicks.[107] In what Patterson calls "translative imitation," Koko translates spoken English into sign, partially or completely. This strongly suggests she understands that a word in English and a word in sign mean the same thing. If Patterson either says or signs BIG GLASS MILK, for example, Koko might imitate in any of five ways. She might repeat exactly: BIG GLASS MILK. Or expand: BIG GLASS MILK DRINK. She may paraphrase: GLASS MILK DRINK. Or reduce: BIG MILK. Or reduce and add a word: BIG MILK DRINK.[108]

Pretense and Deception

Using a fat rubber tube as a trunk, Koko has pretended to be an elephant.[109] Remember Professor Watson's twelve-step scale to measure pretending in children from the previous chapter. Pretending to be an elephant put Koko on Step 1. By pressing a plastic alligator to her nipple and signing "DRINK," she reached Step 2.[110] Koko asked to play with her orangutan doll. She cradled and kissed it and signed

"DRINK." Patterson asked, "Where does the baby drink?" Koko molded the doll's hands into "DRINK MOUTH" and pointed the doll's hands to its mouth. Koko then pointed to the doll's mouth and offered her to Patterson, who said, "The baby said 'drink mouth.' That's right. But where on you?" Koko then touched the doll to her breast and signed "TOILET NIPPLE DRINK" and "NIPPLE DRINK."[111] This was Step 5, one step above the minimum for mental representation.

Comparative psychologist Robert Mitchell argues that deception hinges upon a recognition of regularities in another's response to one's deceptive behavior, whether it's inhibition of behavior, distraction of attention, concealment, or something else. Someone is deceived when she thinks something is true that's not and responds (or doesn't) as if it were.[112] When Mitchell canvassed the literature of deception, he discovered that chimpanzees were reported to use fifty-four types, gorillas forty-one, bonobos thirty-five, and orangutans thirty-four.[113] Here are some examples.

In wild gorilla groups, dominant silverback males usually get the girls. Worse for younger guys, gorillas usually vocalize when copulating. If a nondominant gorilla tries to sneak a copulation when a dominant is in earshot, he usually gives the game away. Things turn ugly. As a result, some nondominants learn to suppress these vocalizations.[114]

At the San Francisco Zoo, Kubie understood that Zura, the object of his iconic gestures, was, Tanner said, "an intentional and responding being."[115] But Zura was more. Both gorillas made "playfaces," involuntary facial expressions that invite play. Apparently, because Bwana (Kubie's father) would break up their interactions if he caught them playing or mating, Zura sometimes used one or both hands to hide her playface from Kubie's sight for several seconds. This usually delayed or prevented play.[116]

Dan Shillito studied gorillas at the National Zoo in Washington, D.C. To Shillito's dismay, one female, Mandara, seemed to dislike him, and often threat-barked. One day, her behavior suddenly changed and she reached through her bars in a friendly flat palm-up beckoning manner. Pleased, Shillito responded. He failed to notice that she was concealing her other hand behind her back and that it held a sharp piece of bamboo. She tried to stab him in the thigh. He escaped injury only when alerted by another scientist at the last moment.[117]

Frans de Waal relates how a gorilla captured her keeper. The gorilla's arm appeared to be stuck in the cage bars. When the keeper saw her struggling, he raced to help. While he was opening the cage, the gorilla hid behind a door and simply wrapped her arms about him when he entered.[118]

Koko has hidden an inside-only toy under her arm so that she could sneak it outside, and has concealed an outside-only stick behind her back to sneak it inside. When asked to throw a toy into her room before going outside, she's made a persuasive throwing motion toward her room without letting the toy go.[119] Christine Tam, one of Patterson's assistants, reports that in 1996, Koko and Michael used signs to communicate their desires to eat, drink, and play. Ndume, who has not been taught signs, usually knocks on the wall of his room or claps for attention. In the morning, he typically knocks to request browse. Koko does not receive browse in the morning. Ndume is also the only gorilla to push out a blanket to "exchange" for food.

One morning, Tam heard knocking coming from Ndume's room. When she saw a towel being pushed out, she held out greens and asked Ndume if he wanted them. There was no response. Tam peered through the window but could see only a gorilla profile with a turned-away face. Then she saw Ndume. He was playing outside. Tam took a closer look inside Ndume's room. It was Koko. She had been imitating Ndume's usual behavior, apparently to get browse when she usually doesn't.[120]

When a bored and ornery Koko signed "DRINK" by sticking her thumb in her ear or, when asked to smile for a photograph, signed, "SAD-FROWN," she illustrated Lyn Miles's Step 4 intentional deception.[121] Caught trying to poke a hole in a screen with a purloined chopstick, Patterson asked Koko what she was doing. Koko stuck the chopstick into her mouth like a cigarette and replied, "SMOKE

MOUTH." On another occasion, when she was chastised for eating a red crayon, she began "applying" the crayon to her lips, as if it were lipstick.[122] These were at Step 5.

Theory of Mind

There is some evidence that gorillas realize that other gorillas have minds and use this knowledge to predict their behavior; that is, they have some aspects of a "theory of mind." Zura's concealment of her playface suggests she could "see" Kubie's mental perspective. She knew she was making a playface and that if Kubie saw it, he would play with her.[123] Scientists sometimes refer to this as an "explicit" theory of mind. It's an ability that human children don't often acquire until their third year, and not in full until their fifth.[124] Juan Carlos Gómez thought that infant Muni couldn't mentally represent. Yet she took his hand to request help in reaching the forbidden latch, alternated eye gazes, and tried to get his attention to communicate. This suggested that she had some sort of theory of mind. Gómez argues that it was "implicit," the same sort of understanding that lay behind her realization that his attention and actions were linked and allowed her to intentionally communicate.[125]

Andrew Whiten watched Delilah, then eight years old, interact with her son, Daniel, at the Bristol Zoo. She sometimes encouraged him to succeed in what he was failing to do: crawling, climbing, or reaching. She supported him when he succeeded and lured him forward, while adjusting her encouragement as needed.[126] Delilah was "scaffolding" her son.

Scaffolding occurs when an adult encourages a child to attain a goal too difficult to reach at his level of development. Adults do things that are impossible for the youngster. This allows him to accomplish what he can.[127] Scaffolding may or may not amount to teaching if teaching requires the intention to educate.[128] Little of what human parents do with infants qualifies as teaching, although it is quite common for an adult simply to place a child into a situation that will assist him in learning.[129] Whiten argues that Delilah recognized that Daniel was trying and failing and that she intentionally helped.[130] He also thinks she might have understood that Daniel did not know certain things he ought to have known, for example, that he shouldn't crawl off the four-foot platform in the enclosure. After Daniel fell off several times, Delilah often dangled him by one arm over the spot where he kept tumbling away and swung him gently.[131]

When Koko molded the hands of her orangutan doll into "DRINK MOUTH," she showed elements of a theory of mind. And she occasionally teaches by correction. When Koko was four, a teacher tried to attract her attention as she gazed out of a window by saying "chicken." Koko, without moving her gaze, signed "NO, GORILLA." Overhearing a conversation in which one human asked another whether Koko was an adolescent and the other replied, "No, she's not an adolescent yet; she's a juvenile," Koko interjected, "NO, GORILLA." When Ron Cohn said, "Well, dynamite," as Koko sat with a red blanket, Koko responded, "NO, RED." At the age of four, she pointed to squash on her plate. The teacher said "Potato." Koko said, "WRONG SQUASH," then pointing, "DO HURRY THAT" and "SQUASH EAT HURRY."[132]

Does Koko Have Practical Autonomy?

Patterson writes,

> By the end of her second year, Koko was able to form mental representations, to use simple skills, to answer basic questions, and to understand symbols. Moreover, Koko readily attained the sensorimotor skills most often depicted as precursors or prerequisites for language. Finally, probably the most compelling evidence of Koko's cognitive development was her acquisition and use of numerous signs. Clearly Koko has the ability to keep some representation of dozens of signs in her mem-

> ory along with some understanding of how to use these signs. . . . At least during the early years of Koko's life, both her mental abilities and her understanding of the world do not appear to differ fundamentally from those of young children.[133]

Koko passed a standard MSR test when she was nineteen.[134] As we saw earlier, she is an experienced and confident mirror user. Like Chantek, she possesses both James's "Me-self" and "I-self," Damasio's autobiographical self, and not just Neisser's ecological and interpersonal selves, but a conceptual self, probably a temporally extended self, and perhaps a private self. Koko hints at the last. Conceding that it is not easily quantified, Patterson and others who studied Koko and Michael detect an ability to engage "in some sense" in narrative or storytelling. Michael, for example, told his teacher about an argument he had heard; Koko participated in a story in which she helped provide motives and emotions for the characters; and Michael complained that he received a snack of stew, and not his usual peanut butter sandwich, the day before when he was ill, and described events the researchers believe occurred when he was captured in Africa as an infant.[135]

Koko consistently scored between 70 and 95 on standard human child intelligence tests. Gorillas undoubtedly surmount Piagetian Substage 4 tests for object permanence.[136] A twenty-two-month-old female gorilla born in the Rome Zoo passed Substage 6.[137] When Koko was twenty-one months, she passed, too.[138] This means that she can solve problems by insight. She pretends at a relatively high level. All this indicates that she clearly mentally represents. She imitates at the action level. Her language involves hundreds of signs, and she modulates her use of them in at least ten ways. She understands several thousand English words. She intentionally communicates, intentionally deceives, occasionally teaches through correction, and shows elements of a theory of mind. Koko is clearly a Category One animal. Because her mental abilities appear a tad more sophisticated than Chantek's, a tad less than Kanzi's (the bonobo whom I discussed in Chapter 3), I assign her an autonomy value of 0.95.

TWELVE

Legal Rights for Nonhuman Animals

Dignity-Rights

Any being who can want, who can intentionally act to get what she wants, and who has a sense of self that allows her to know it is she who wants and she who is acting to get what she wants possesses practical autonomy sufficient for dignity-rights. Her species is irrelevant.

Category One animals, whose autonomy values lie between 0.90 and 1.0., clearly possess practical autonomy sufficient for dignity-rights. Christopher achieved it around his first birthday. Koko the gorilla (0.95), Chantek the orangutan (0.93), and the Atlantic bottle-nosed dolphins, Ake and Phoenix (0.90), are also Category One animals. In Category Two are animals whose autonomy values range from 0.51 to 0.89. Here we find Alex the African Grey parrot (0.78), Echo the African elephant (0.75), Marbury the dog (0.68), and your average honeybee (0.59).

In Chapter 3, I proposed a moderate use of the precautionary principle, the principle emerging from environmental law that says we should act to prevent threatened harm some if evidence, if not proof, of harm exists. Any animal with an autonomy value higher than 0.70 would be presumed to have practical autonomy sufficient for dignity-rights. Alex and Echo thus join apes and dolphins as dignity-rights-holders. I also proposed that a moderate reading of the precautionary principle should have us strongly consider granting some proportional dignity-rights, even to an animal with an autonomy value of 0.65, perhaps even 0.60. If this argument is accepted, Marbury will enter the circle of legal persons as a proportional rights-holder. Only an expansive, and unlikely, interpretation of the precautionary principle would include a honeybee.

Basic Equality Rights

We have seen that the mental abilities of Alex, Echo, Phoenix and Ake, Chantek, Koko, and Christopher are sufficient to entitle them to dignity-rights. The same data can also be used to determine their entitlement to equality. Liberty means you receive rights because of what you are, without being compared to anyone. Equality is different. It demands a comparison. Because likes should be treated alike, something can be equal only *to* something or to someone else. In 1935, Oklahoma enacted the Habitual Criminal Sterilization Act. Anyone convicted three times of a felony involving moral turpitude would be sterilized, save for moonshiners, tax cheats, political offenders, and embezzlers. One Skinner, twice convicted of robbing with firearms and once for stealing chickens, argued to the United States Supreme Court that the Sterilization Act violated his right to equality under the Fourteenth Amendment to the United States Constitution. The judges agreed.[1]

The Constitution's equality guarantee "does not require different things which are different in fact . . . to be treated in law as though they were the same."[2] But the Sterilization Act carried beyond. Larceny of $25 brought the scalpel, but not embezzlement of the same amount.[3] The problem is that embezzlement and larceny are nearly the same thing, their differences turning on arcane historical distinctions that few cared about in the twentieth century. You commit larceny by carrying away someone else's property with the intent to deprive him of it at the time you take it. You embezzle when you fraudulently convert someone else's property that was entrusted to you. The bank robber commits larceny; the bank president embezzles. The Supreme Court said, "When the law lays an unequal hand on those who have committed intrinsically the same quality of offense and sterilizes one and not the other, it has made as invidious a discrimination as if it had selected a particular race or nationality for oppressive treatment. . . . [T]he equal protection clause would indeed be a formula of empty words if such conspicuously artificial lines could be drawn."[4]

Equality, the idea that likes should be treated alike, is among the highest values and principles of Western law.[5] It has a glorious history. In law professor Peter Westen's opinion, "some of the most significant and acclaimed moral crusades in history have been fought and won under the banner of 'equality,' including the abolition of chattel slavery, the elimination of feudal privilege, the outlawing of caste, the disestablishment of religion, the spread of universal suffrage, the opening of careers to talent, the outlawing of racial discrimination and the emancipation of women."[6]

Because courts and legislatures are forever making distinctions and drawing lines, and necessarily so, equality can be hard to apply.[7] Early in its history, the European Court of Human Rights turned back the argument that every difference in treatment violates the European Convention on Human Rights.[8] European discrimination is legal only if some objective and reasonable justification exists. It must further some legitimate aim, and a reasonable relationship must exist between that aim and the means employed to attain it.[9] This is essentially how the United States Supreme Court normally applies the equality provision of the Constitution.[10] But both European and American judges closely scrutinize distinctions based on race, sex, or illegitimacy.[11] For example, rejecting a state military academy's exclusion of women, the United States Supreme Court said states could "not exclude qualified individuals based on 'fixed notions concerning the roles and abilities of males and females.'"[12]

A Realizable Minimum

In 1858, Abraham Lincoln and Stephen A. Douglas locked in a monumental series of debates across the Illinois prairie to decide a seat in the United States Senate. Douglas tarred Lincoln an "abolitionist," committed to "perfect equality" between whites and blacks. In pre–Civil War Illinois, this was like Joseph McCarthy's happening to mention in 1952 that you were a Communist.[13] The American government, Douglas insisted, had been "made by white men for the benefit of white men and their posterity forever."[14] By his day's standards, Lincoln was a moderately enlightened thinker about black chattel slavery. He always opposed it, even if he did not always support social and political equality for blacks. But he made it the central issue of his Senate campaign, accentuating "the difference between those who think it wrong and those who do not think it wrong," and squarely traced it to the assertion of the Declaration of Independence that "all men are created equal."[15] That was fine with Douglas, for it allowed him to press the attack. Lincoln, he exclaimed, was a radical, committed not just to freedom for the slave, but to the complete social and political equality of white and black. Lincoln correctly sensed that if Douglas succeeded, he was lost, and edged toward the center.

Lincoln called for slavery's end, while attempting to dodge the stones of "perfect equality" that Douglas kept hurling.[16] During their fourth debate, which took place at Charleston, Lincoln argued against the view that "because the white man is to have the superior position, that it requires that the negro should be denied everything. I do not perceive that because I do not court a negro woman for a slave that I must necessarily want her for a wife. My understanding is that I can just leave her alone."[17] In the sixth debate, at Quincy, he was even more explicit: "I have no purpose to introduce political and social equality between the white and the black races . . . but . . . there is no reason in the world why the negro is not entitled to all the natural rights enumerated in the Declaration of Independence—the right to life, liberty, and the pursuit of happiness."[18]

Historians have made clear what Lincoln was trying to do. David Zarefsky argues that he avoided "the slippery slope by which freedom led to racial equality" by declaring freedom an economic right that did not necessarily carry social and political equality with it.[19] David Potter labeled Lincoln's the "minimum anti-slavery position," while Garry Wills said Lincoln's "nub, the realizable minimum," was that "at the very least, it was wrong to treat human beings as property."[20] Lincoln was a famously practical lawyer, president, and commander in chief, known in the courtroom, political arena, and war room for affably conceding one nonessential point after another, until his opponent believed he had won. But Lincoln rarely allowed the essential to slip away.

Obtaining any legal rights for nonhuman animals in the present legal system requires fighting from the platform of Lincoln's realizable minimum. Lincoln believed the physical, historical, legal, religious, economic, political, and psychological realities of his day meant that taking more than one step at a time for black slaves would lead to no change in their legal status. In the 1850s, that meant that advocating the social and political equality of black slaves, whatever Lincoln personally believed, would result in their continued enslavement. Today, it means that advocating for too many rights for too many nonhuman animals will lead to no nonhuman animal's attaining rights.

Comparing Human and Nonhuman Animals

In September 2000, physiology professor Colin Blakemore, Britain's most outspoken proponent of the use of all nonhuman animals in biomedical research, told the Fifth International Congress on Bioethics that humans must use apes and monkeys in biomedical research because they are so like us. Why? Because "[r]esearch on a species which is similar to humans is more likely to generate results which are relevant. It is a dilemma that we all have to acknowledge."[21] The similarities to which Blakemore refers are mental, for what dilemma exists if the foot of a nonhuman animal is similar to a human foot, or the hand, hair, or eye? When one hears the word "similarity," one should think about equality.

The strongest argument for equality rights incorporates the argument for dignity-rights and goes like this: a normal human child possesses the practical autonomy sufficient for dignity-rights by age eight months, perhaps by four months. Any nonhuman animal with practical autonomy is similar to this child in ways highly relevant to the possession of basic legal rights. As a matter of equality, Koko is certainly entitled to them. In many ways, her mind resembles the mind of a toddler, even that of a preschooler. She has a "Me-self" and an "I-self," an autobiographical self, ecological, interpersonal, and conceptual selves, probably a temporally extended self, and perhaps a private self. She can pass a mirror self-recognition test and a Substage 6 object permanence test, imitate at the action level, form complex mental representations, understand hundreds of signs, answer basic questions, score between 70 and 95 on standard human child intelligence tests, intentionally communicate and deceive, occasionally teach through correction, and demonstrate elements of a theory of mind. The minds of Kanzi the bonobo and Washoe the chimpanzee, whom we met in *Rattling the Cage,* may even be a bit more sophisticated, while Chantek the orangutan's and the Atlantic bottle-nosed dolphins, Phoenix and Ake, minds may be just a bit less complicated. But every Category One animal has a strong equality argument for basic rights.

As the minds of nonhuman animals resemble less and less the minds of human preschoolers, toddlers, and infants, either because they become simpler or just different—Alex to Echo to Marbury to honeybees—the argument for quality rights weakens. It may still remain strong, just not as strong. But at some point, our minds are no longer sufficiently alike to trigger equality rights at all. Where that point lies is no clearer for equality than it is for liberty, but a moderate application of the precautionary principle suggests generosity.

In Chapter 3, I noted that at some point, autonomy disappears and with it any nonarbitrary entitlement to liberty rights on the ground of possessing practical autonomy. But judges and legislators might still decide to grant even a completely nonautonomous being basic rights. To the extent they confer rights arbitrarily, the argument that as a matter of equality, nonautonomous animals of many species should be entitled to basic rights is strengthened, too.

Judges and legislators actually do this. Some humans have little or no autonomy but have legal rights. We were introduced to a couple of them in *Rattling the Cage:* Joseph Saikewicz, a sixty-seven-year-old man with an IQ of ten, and Beth, a ten-month-old girl born into a permanent vegetative state.[22] The state of Louisiana has even enacted a statute that designates a fertilized *in vitro* ovum a legal person before it is implanted in a womb.[23] Louisiana judges may appoint curators to protect its rights, and the fertilized ovum can even sue and be sued.[24]

In light of equality's demand that likes be treated alike, a modern Stephen Douglas might argue that every nonhuman animal with the cognitive abilities of a Beth, even a fertilized ovum, should automatically be entitled to the rights to which these humans are entitled. The purpose of anyone's advancing such an argument would probably not be to gain rights for a large number of nonhuman animals but to frighten judges and legislators away from taking the first, and most obvious, steps to grant legal rights to any nonhuman animal.

Imagine that a little girl like Beth was born in a local hospital to a young woman who doesn't want her. No one knows who the father is. Her nearly nonexistent brain does nothing more than keep her alive. Her heart pumps, her lungs inflate, her intestines absorb. But she is neither conscious nor sentient. She cannot think or feel. She is so utterly devoid of any higher brain functions that emergency surgery was performed on her immediately after birth without anesthesia; she didn't need any. She has no mind. Yet she has the legal right to bodily integrity, though it must be exercised by someone else. It is important to note that no one suggests she be eaten or used in terminal biomedical research.

The bestowal of rights upon a Joseph Saikewicz, a Beth, or a fertilized embryo strengthens the argument for equality rights not just for Category One animals but for Category Two animals as well. Compare our baby girl to Koko. On what nonarbitrary ground could a judge find the little girl has a common law right to bodily integrity that forbids her use in terminal biomedical research, but that Koko shouldn't have that right, without violating basic notions of equality? Only a radical speciesist could accept a baby girl who lacks consciousness, sentience, even a brain, as having legal rights just because she's human, yet the thinkingest, talkingest, feelingest apes have no rights at all, just because they're not human.[25]

It is the extreme disparity, the utter arbitrariness of the distinction that powers the argument for equality rights for all Category One and some Category Two animals. This distinction is so extreme, so arbitrary, that it obviously violates the principle of equality at its most fundamental level. The disparity decreases as an animal's autonomy value lessens. As a value approaches 0.50, the disparities become small enough to allow a judge to distinguish rationally between that creature and a severely retarded man. And at some point, the psychological and political barriers to equality for a nonhuman animal with a low autonomy value become insuperable.

Overcoming Evil

Of wild Chanteks (orangutans), Gisela Kaplan and Lesley Rogers have written that they "are too large to keep in enclosures, too intelligent to keep in zoos, too self-aware to keep in laboratories, too sensitive to be exploited in shows and circuses, and too close to us to ignore the fact that they too have a right to live freely."[26] Bodily integrity is the most fundamental right. Bodily liberty is just as important, at least for mammals. These rights would prevent orangutans from being exploited in the ways that Kaplan and Rogers point out. They would also prevent the capture of a Phoenix or Ake from the Gulf of Mexico, the culling of an Echo at Amboseli, the killing of a wild Koko for bushmeat in Rwanda, and the capture of a wild Alex in Africa for the pet trade. Eighteenth-century British prime minister Lord Shelburne believed that "it requires no small labor to open the eyes of either the public or an individual but, when that is accomplished you are not got a third of the way. The real difficulty remains in getting people to apply the principles which they have admitted and of which they are now so fully convinced. Then springs the mine composed of private interests and personal animosity."[27] By continuing to expose the arbitrariness and bias that girds the historical failure to grant basic rights for even the most deserving nonhuman animals, this book has taken you along Shelburne's first third of the way in the struggle to attain these rights.

Human slavery was once as firmly entrenched as nonhuman animal slavery is today. Eighteenth-century revolutionary Virginia produced the great American founders George Washington, Thomas Jefferson, James Madison, and George Mason. They were instrumental to constructing a nation self-consciously dedicated to liberty and equality. Yet they "were slaveholders, and substantial ones," slavery historian Ira Berlin has written. "At one time or another, each condemned slavery as evil and recognized the contradiction of his own stated beliefs to the realities of chattel bondage. None, however, had the courage to translate his intellectual commitment to universal freedom into public policy or even individual action."[28]

The obstacles to basic legal rights for any nonhuman animal, physical, economic, political, religious, historical, legal, and psychological, that I set out in Chapter 2 are major and real. But despite the blindness of even our greatest citizens, these obstacles have been hurdled before, at last with regard to humans. The most significant example is the abolition of human slavery, within decades after the American

Revolution. "Considering that slavery had been globally accepted for millennia, it is encouraging that people were able to make such a major shift in their moral view, especially when a cause like abolition conflicted with strong economic interests," David Brion Davis has written. "We can still learn from history the invaluable lesson that an enormously powerful and profitable evil can be overcome."[29] Then we will have taken the first and most crucial step toward unlocking the cage. Judges must recognize that even using a human yardstick, at least some nonhuman animals are entitled to recognition as legal persons.

Who Should Get Basic Liberty Rights under the Common Law?

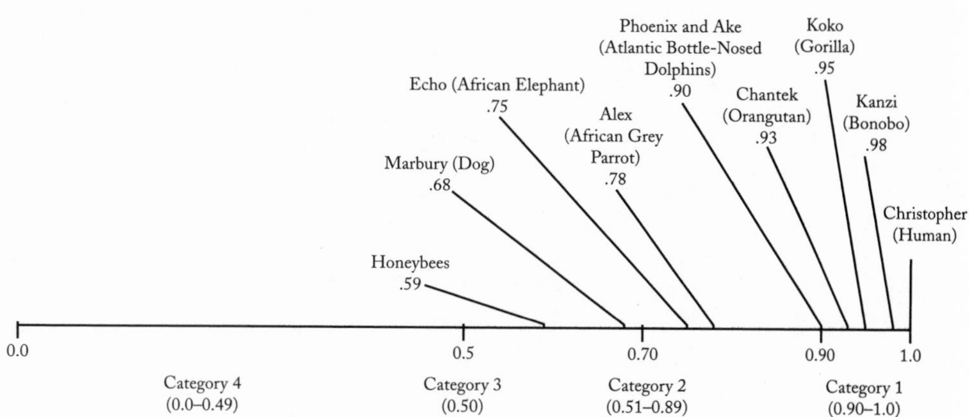

Autonomy Values

Category 1 = Nonhuman animals who clearly possess sufficient autonomy for basic liberty rights. An autonomy value of .90 is the cut-off for basic legal rights using a narrow reading of the precautionary principle.

Category 2 = Nonhuman animals who, according to increasing evidence, possess sufficient autonomy for basic liberty rights. An autonomy value of .70 is the cutoff for basic liberty rights using a moderate reading of the precautionary principle.

Category 3 = Nonhuman animals about whom we do not know enough to reasonably determine whether they possess sufficient autonomy for basic liberty rights. This category probably includes most species.

Category 4 = Nonhuman animals who, according to increasing evidence, lack sufficient autonomy for basic liberty rights.

NOTES

Chapter 1 In a Ugandan Rain Forest

1. Michael Tomasello, *The Cultural Origins of Human Cognition* 54 (Harvard University Press 1999). "Nonhuman primates definitely do have an understanding of all kinds of complex physical and social events, they possess and use many kinds of concepts and cognitive representations, they clearly differentiate between animate and inanimate objects, and they employ in their interactions many complex and insightful problem-solving strategies," *id.* at 19.

2. *See generally,* Frans de Waal, *The Ape and the Sushi Master* (Basic Books 2001).

3. Marc Hauser, *Wild Minds: What Animals Really Think* 58–59 (Henry Holt 2000). Hauser theorizes that instead of counting, chimpanzees may use a "memory storage system."

4. Nobuyuki Kawai and Tetsuro Matsuzawa, "Numerical memory span in a chimpanzee," 403 *Nature* 39 (2000).

5. Charles A. Menzel, "Unprompted recall and reporting of hidden objects by a chimpanzee (Pan troglodytes) after extended delay," 113(4) *Journal of Comparative Psychology* 426 (1999).

6. Christopher Boehm, *Hierarchy in the Forest: The Evolution of Egalitarian Behavior* 171–172, 180, 181 (Harvard University Press 1999).

7. *Id.* at 182.

8. Brian Hare *et al.*, "Chimpanzees know what conspecifics do and do not see," 59(4) *Animal Behavior* 771 (2000).

9. Personal communication from Brian Hare, dated August 10, 2001.

10. Brian Hare *et al.*, "Do chimpanzees know what conspecifics know?" 61 *Animal Behavior* 139 (2001).

11. Richard Wrangham and Dale Peterson, *Demonic Males: Apes and the Origins of Human Violence* 253–254 (Mariner Books 1996).

12. *Id.* at 255.

13. Saint Augustine, *The City of God,* Book IV, sec. 4, at 112 (George Wilson, trans., The Modern Library 1950).

Chapter 2 One Step at a Time

1. William Lee Miller, *Arguing About Slavery: The Great Battle in the United States Congress* 11 (Knopf 1996).

2. *See generally,* Hugh Thomas, *The Slave Trade: The Story of the Atlantic Slave Trade 1440–1870* (Simon & Schuster 1997).

3. David Brion Davis, "The enduring legacy of the South's Civil War victory," *New York Times* (August 26, 2001).

4. Verlyn Klinkenborg, "Cow parts," 22 *Discover* 52, 53–62 (August 2001).

5. William Lee Miller, *supra* note 1, at 11.

6. Richard A. Epstein, "The next rights revolution," *National Review* 44, 45 (November 8, 1999).

7. Hugh Thomas, *supra* note 2, at 515.

8. Caryl Phillips, *The Atlantic Sound* 33–34 (Faber and Faber 2000).

9. Hugh Thomas, *supra* note 2, at 506, 508, 513, 525, 528, 550, 554.

10. *Id.* at 515, 516.

11. *Id.* at 516, 540.

12. Thomas Roderick Dew, "Abolition of negro slavery," 12 *American Quarterly Review* 189 (1832), reprinted in *The Ideology of Slavery: Proslavery Thought in the Antibellum South, 1830–1860* 21, 30 (Drew Gilpin Faust, ed., Louisiana University Press 1981).

13. Hugh Thomas, *supra* note 2, at 525, 530, 540.

14. David Brion Davis, *The Problem of Slavery in the Age of Revolution 1770–1823* 423 (Cornell University Press 1975); John Stoughton, *William Wilberforce* 66–67 (A. C. Armstrong & Son 1880).

15. Hugh Thomas, *supra* note 2, at 540.

16. *Id.* at 13.

17. Laurence H. Tribe, Presentation at Faneuil Hall, Boston, Massachusetts, February 8, 2000, at 48:40 (videotape on file with author).

18. XV *Oxford English Dictionary* 665 ("Slave," def. 1, at 665 [2d ed. 1989]).

19. Slavery Convention, 60 LNTS 253, Article 1 (1926).

20. Isaiah Berlin, "Two concepts of liberty," in *The Proper Study of Mankind* 194 (Henry Hardy and Roger Hausheer, eds., Farrar, Straus & Giroux 1997).

21. Orlando Patterson, *Slavery and Social Death: A Comparative Study* 7 (Harvard University Press 1982).

22. David Brion Davis, S*lavery and Human Progress* 323 note 14 (Oxford University Press 1984).

23. Alan Watson, "Rights of slaves and other owned-animals," 3 *Animal Law* 1 (1997).

24. Edith Hamilton, *The Echo of Greece* 23 (W. W. Norton & Co. 1957).

25. Hugh Thomas, *supra* note 2, at 237, 438.

26. Charles Darwin, *The Voyage of the Beagle* 502 (P. F. Collier & Son 1909) (1845).

27. *Id.*

28. *See generally,* Orlando Patterson, *Freedom in the Making of Western Culture* (Basic Books 1991).

29. Charles Darwin, *supra* note 26, at 502.

30. Richard Sorabji, *Animal Minds and Human Morals: The Origin of the Western Debate* 219 (Cornell University Press 1993).

31. David Brion Davis, *The Problem of Slavery in Western Culture* 69 (Cornell University Press 1966).

32. Orlando Patterson, *supra* note 21, at 7, quoting *Grundisse* 325–326 (Martin Nicolaus, trans., Penguin and New Left Books 1973) (emphasis in the original).

33. Frederick Douglass, *Narrative of the Life of Frederick Douglass: An American Slave Written by Himself* 74 (Benjamin Quarles, ed., Harvard University Press 1960).

34. Alice Walker, "Preface," in Marjorie Spiegel, *The Dreaded Comparison: Human and Animal Slavery* 9 (New Society Publishers 1988).

35. Douglas Chadwick, *The Fate of the Elephant* 225 (Sierra Club Books 1992).

36. *Id.* at 13 (emphasis in the original).

37. IV Roscoe Pound, *Jurisprudence* 192 (West Publishing Co. 1959) (The constitution of the emperor Antonius Pius gave slaves power to bring suits to determine whether they were being cruelly treated, but not to obtain their freedom, *id.* at 192–193). *See* "Introduction" to *The Institutes of Justinian* 27 (Thomas Collett Sanders, trans., William S. Hein and Co. 1984) (1876).

38. Barry Nichols, *An Introduction to Roman Law* 69 (Oxford University Press 1962). *See* W. W. Buckland, *A Manual of Roman Private Law* 37 (2d ed., Cambridge University Press 1939).

39. *Id.*; *see* John Chipman Gray, *The Nature and Sources of the Law* 43 (2d ed., The Macmillan Co. 1931).

40. A. Leon Higginbotham, "Property first, humanity second: The recognition of the slave's human nature in Virginia slave law," 50 *Ohio State Law Journal* 511, 512–514 (1989). *See Peter v. Hargrave,* 46 Va. (5 Gratt) 12, 17 (1848).

41. *See George (A Slave) v. The State,* 37 Miss. 316 (1859).

42. Richard D. Ryder, *Animal Revolution: Changing Attitudes Toward Speciesism* 83, 89–90 (Basil Blackwell 1989).

43. Zulma Steele, *Angel in Top Hat* 1–11, 195–205 (Harpers and Brothers Publishers 1942).

44. Ellen Carol DuBois, *Feminism and Suffrage: The Emergence of an Independent Women's Movement in America, 1848–1869* 31–39, 51 (Cornell University Press 1978).

45. Peter Singer, *Animal Liberation* 221 (2d ed., New York Review Book 1990).

46. *See* Orlando Patterson, *supra* note 21 (discusses how slavery operated in sixty-six societies).

47. Genesis 1:28.

48. I explain this in Steven M. Wise, "The legal thinghood of nonhuman animals," 23 *Boston College Environmental Affairs Law Review* 471, 480–489 (1996); Steven M. Wise, "How nonhuman animals were trapped in a nonexistent universe," 1 *Animal Law* 15, 30–34 (1995).

49. Robert Speth, Ph.D., review of *Rattling the Cage: Toward Legal Rights for Animals* 6(2) *Newsletter of the Society for Veterinary Ethics* 6, 7 (May 2000).

50. Louis Henkin, *The Age of Rights* x (Columbia University Press 1990) (emphasis added). *See id* at 183–188.

51. David Brion Davis, *supra* note 22, at 82–101; Orlando Patterson, *supra* note 21, at 40–41, 117.

52. Ronald Segal, *Islam's Black Slaves: The Other Black Diaspora* (Farrar, Straus & Giroux 2001); Hugh Thomas, *supra* note 2, at 37, 298; *see id.* at 29–31; Orlando Patterson, *supra* note 21, at 117. *See also* David Brion Davis, *supra* note 22, at 60.

53. Thornton Stringfellow, "A brief examination of scripture testimony on the institution of slavery," reprinted in part in Drew Gilpin Faust, ed., *supra* note 12, at 138; Thornton Stringfellow, *Scriptural and Statistical Views in Favor of Slavery* 105 (J. W. Randolph 1856). *See* Thornton Stringfellow, "The Bible argument, or, slavery in the light of divine revelation," in *Cotton Is King and Pro-Slavery Arguments* 459 (Pritchard, Abbott & Loomis 1860).

54. David Brion Davis, *supra* note 22, at 112.

55. *E.g.*, Nancy Tuana, *The Less Noble Sex: Scientific, Religious, and Philosophical Conceptions of Woman's Nature* 10–14, 157–159 (University of Indiana Press 1993).

56. Anonymous, *The Lawes Resolutions of Women's Rights: Or, the Lawes Provisions for Women* 125 (1632). *See* Eleanor Flexnor, *Century of Struggle: The Woman's Rights Movement in the United States* 7–8 (Atheneum 1973).

57. Dava Sobel, *Galileo's Daughter* (275–276) (Random House 1999).

58. Pam Belluck, "Board decision on evolution roils an election in Kansas," *New York Times,* at A1, A10 (July 29, 2000).

59. Ernst Mayr, "Darwin's influence on modern thought," 283 *Scientific American* 79, 83 (July 2000).

60. David Baltimore, "50,000 genes, and we know them all (almost)," *New York Times,* at D17 (June 25, 2000).

61. *E.g., Thomas v. Review Board,* 450 U.S. 707, 714 (1981) (interpreting the religious liberty clause of the First Amendment to the United States Constitution).

62. *Id.* at 2, 195–196.

63. Steven M. Wise, *supra* note 48, at 492–538; Steven M. Wise, *supra* note 48, at 21–31.

64. Steven M. Wise, *supra* note 48, at 34–43.

65. Richard Sorabji, *supra* note 30, 20–21, 42, 58–61.

66. *Id.* at 52, quoting Seneca, *Ep.* 124.16.

67. Steven M. Wise, *supra* note 48, at 18–31.

68. Richard Sorabji, *supra* note 30, at 199; Keith Thomas, *Man and the Natural World* 19–20 (Pantheon Books 1983).

69. C. S. Lewis, *Vivisection: A Rational Discussion* 7 (New England Anti-Vivisection Society undated).

70. Roscoe Pound, *supra* note 37, at 194, 529, 530. *See* Carleton Kemp Allen, "Things," 28 *California Law Review* 421, 422 (1940); Thomas Collett Sanders, trans., *supra* note 37, at 26.

71. Roscoe Pound, *supra* note 37, at 192.

72. David Brion Davis, *supra* note 31, at 58; Alan Watson, *Roman Slave Law* 46 (Johns Hopkins University Press 1987) (In Rome, a slave "always remained . . . corporeal property whose value could be measured in monetary terms").

73. John Austin, *Lectures on Jurisprudence* 368 (4th ed., Robert Campbell, ed., John Murray 1873).

74. P. A. Fitzgerald, *Salmond on Jurisprudence* 298 (12th ed., Sweet and Maxwell 1966).

75. Daniel Defoe, *The Kentish Petition,* addenda 1.11 (1701).

76. Note, "What we talk about when we talk about persons: The language of a legal fiction," 114 *Harvard Law Review* 1745, 1746 (2001).

77. Barbara Tuchman, "Why policy-makers don't listen," in *Practicing History: Selected Essays* 287 (Knopf 1981); Stephen R. L. Clark, *The Moral Status of Animals* 7 (Oxford

University Press 1984); Thomas S. Kuhn, *The Structure of Scientific Revolutions* 7 (2d ed., University of Chicago Press 1963).

78. Norwood R. Hanson, *Patterns of Discovery: An Inquiry into the Conceptual Foundations of Science* 4–30 (Cambridge University Press 1965).

79. Oliver Wendell Holmes, "Ideals and Doubts," 19 *Illinois Law Review* 1, 2 (1915).

80. Sandea Blakeslee, "Watching how the brain works as it weighs a moral dilemma," *New York Times,* at D3 (2001); Joshua D. Greene *et al.*, "An MRI investigation of emotional engagement in moral judgment," 293 *Science* 2105 (2001).

81. *See* 1 Laurence H. Tribe, *American Constitutional Law,* sec. 1–15–1–17, at 70–89 (Foundation Press 2000); Laurence H. Tribe and Michael C. Dorf, *On Reading the Constitution* 65–80 (Harvard University Press 1991) (values in constitutional decisionmaking). *Cf.* P. S. Atiyah and R. S. Summers, *Form and Substance in Anglo-American Law* 411 (Oxford University Press 1987).

82. Oliver Wendell Holmes, Jr., "Natural law," 33 *Harvard Law Review* 40, 41 (1918–1919).

83. *Id. See* Cass Sunstein, "Incommensurability and valuation in law," 92 *Michigan Law Review* 779, 810–811 (1994); Richard Warner, "Incommensurability as a jurisprudential puzzle," 68 *Chicago-Kent Law Review* 147, 168 (1992).

84. David Brion Davis, *supra* note 3.

85. William Lee Miller, *supra* note 1, at 15.

86. C. S. Lewis, *supra* note 69, at 7–8.

87. David Mamet, *Henrietta* 9 (Houghton Mifflin 1999).

88. *Id.* at 4.

89. XVI *Oxford English Dictionary* 157 (2d ed. 1989).

90. R. G. Frey, *Rights, Killing, and Suffering* 115 (Basil Blackwell 1983).

91. Gisela Kaplan and Lesley J. Rogers, *The Orangutans* 31 (Perseus Publishing 2000).

92. Marc D. Hauser, *Wild Minds: What Animals Really Think* 159 (Henry Holt and Co. 2000).

93. Irene Pepperburg, *The Alex Studies: Cognitive and Communicative Abilities of Parrots* 184 (Harvard University Press 1999).

94. Wesley Newcomb Hohfeld, *Fundamental Legal Conceptions as Applied in Judicial Reasoning* 64 (Walter Wheeler Cook, ed., Yale University Press 1919). *See* W. L. Morison, *John Austin* 164 (Edward Arnold 1982); Joseph William Singer, "The legal rights debate in analytical jurisprudence from Bentham to Hohfeld," 1982 *Wisconsin Law Review* 975, 989 note 22; Walter J. Kamba, "Legal theory and Hohfeld's analysis of a legal right," *Jurisprudence Review* 249, 249 (1974).

95. David Lyons, "Correlativity of rights and duties," *4 Nous* 46 (1970).

96. Rex Martin, *A System of Rights* 31 (Oxford University Press 1993).

97. Wesley Newcomb Hohfeld, *supra* note 94, at 38, 39.

98. Steven M. Wise, *Rattling the Cage: Toward Legal Rights for Animals* (Perseus Books 2000).

99. *Virani v. Jerry M. Lewis Truck Parts & Equipment, Inc.,* 89 F. 3d 574, 577 (9th Cir. 1996), quoting Judith Jarvis Thomson, *The Realm of Rights* 9 (Harvard University Press 1990); P. J. Fitzgerald, *Salmond on Jurisprudence* 229 (12th ed., Sweet and Maxwell 1966).

100. Walter J. Kamba, *supra* note 94, at 256.

101. Rex Martin, *supra* note 96 at 3; L. W. Sumner, *The Moral Foundation of Rights* 37–38 (Oxford University Press 1987).

102. Steven M. Wise, *supra* note 98, at 59.

103. *Norway Plains Company v. Boston and Maine Railroad,* 67 Mass. (1 Gray) 263, 267 (1854).

104. Joan Dunayer argues that "companion nonhuman" should be used in preference to "companion animal" or "pet," Joan Dunayer, *Animal Equality: Language and Liberation* 204 note 8 (Ryce Publishing 2001).

105. *Rabideau v. City of Racine,* 238 Wis. 2d 96, 617 N.W. 2d 678 (2001).

106. Robert S. Summers, "Form and substance in legal reasoning," in *Legal Reasoning and Statutory Interpretation* 11 (Jan van Dunne, ed., Gouda Quint 1989).

107. P. S. Atiyah and R. S. Summers, *supra* note 81, at 5.

108. Melvin Aron Eisenberg, *The Nature of the Common Law* 26–37 (Harvard University Press 1988); Robert S. Summers, "Two types of substantive reasons: The core of a theory of common law justification," 63 *Cornell Law Review* 707, 717–718, 722–724 (1978); Harry H. Wellington, "Common law rules and constitutional double standards: some notes on adjudication," 83 *Yale Law Journal* 221, 223–225 (1973).

109. Orlando Patterson, *supra* note 28, at ix (liberty as supreme value); *Legal Consequences for States of the Continued Presence of South Africa in Namibia (South West Africa) Notwithstanding Security Council Resolution* 276 (1970), 1971 I.C.J. 16, 77 (June 21) (separate opinion of Judge Ammoun, Vice-President) (equality the most important); Richard A. Posner, "Legal Reasoning from the top down and from the bottom up: The Question of unenumerated fundamental rights," 59 *University of Chicago Law Review* 433, 434 (1992).

110. Richard A. Posner, "Animal rights," 110 *Yale Law Journal* 527, 527 (2000).

111. *Id.* at 235.

112. *Id.* at 194, 196, 197, 198, 199, 203, 232, 236, 237.

113. Carleton Kemp Allen, *supra* note 70, at 424, quoting Baron, *Pandekten,* sec. 37, quoted in Holland, *Jurisprudence* 103 (13th ed. 1924) (emphases in original). *See* Roscoe Pound, *supra* note 37, at 530–531.

114. *See* Roscoe Pound, *supra* note 37, at 194–199.

115. Ellen Langer, *The Power of Mindful Learning* 4 (Addison Wesley 1997); Barbara Herman, *The Practice of Moral Judgment* 227–228 (Harvard University Press 1993).

116. Immanuel I. Kant, *Groundwork of the Metaphysic of Morals* 114-131 (H. S. Papp, trans., Harper Torchbooks 1964). *See* Barbara Herman, *supra* note 115.

117. Isaiah Berlin, *supra* note 20, at 216. *See* Michael Allen Fox, "Animal experimentation: A philosopher's changing views," in 3 *Between the Species* 260 (Spring 1987).

118. Carl Wellman, *Real Rights* 113–114 (Oxford University Press 1995). *See* Daniel A. Dombrowski, *Babies and Beasts* 45–140 (University of Illinois Press 1997) (discussing many modern philosophers); H. L. A. Hart, "Are there any natural rights?" in *Theories of Rights* 79, 82 (Jeremy Waldron, ed., Oxford University Press 1984); R. G. Frey, *Interests and Rights: The Case Against Animals* 30 (1980). *See also* A. John Simmons, *The Lockean Theory of Rights* 201 note 93 (Princeton University Press 1992) (listing modern philosophers who claim that children cannot have rights because they lack the capacities for agency, rationality, or autonomy); Katherine Hunt Federle, "On the road to reconceiving rights for children: A postfeminist analysis of the capacity principle," 42 *DePaul Law Review* 983, 987–999 (1993) (Hobbes, Locke, Rousseau, Bentham, and Mill).

119. Frans de Waal, "We the people (and other animals . . .)," *New York Times,* at A23 (August 20, 1999); Frans de Waal, *Good Natured: The Origins of Right and Wrong in Humans and Other Animals* 215 (Harvard University Press 1995).

120. "Boy, 6, fatally shoots classmate in Mich. school," *Boston Globe,* at A1, A16 (March 1, 2000).

121. Isaiah Berlin, *supra* note 20, at 219.

122. *E.g., Airedale NHS Trust v. Bland,* 1 All ER 821 (Family Division 1992), *aff.* 1 All ER 833, 843, 848 (Ct. of App. 1993) (Butler-Sloss, LJ); *id.* at 852 (Hoffman, LJ) (both of the Court of Appeals); *aff.*, 1 All ER 858, 862 (HL 1993) (Lord Goff of Chieveley); *id.* at 889 (Lord Mustill). *See* Steven M. Wise, *supra* note 98, at 247.

123. *Airedale NHS Trust, supra* note 122, at 851, 852 (Hoffman, LJ); *id.* at 846 (Butler-Sloss, LJ) (both of the Court of Appeals); *id.* at 866 (Lord Goff of Chieveley).

124. *E.g., Winterwerp v. Netherlands,* A. 33, para. 51 (Eur. Ct. Hum. R. 1979); *O'Connor v. Donaldson,* 422 U.S. 563 (1975).

125. *E.g., Guardianship of Doe,* 583 N.E. 2d 1263, 1268 (Mass. 1992); *id.* at 1272–1273 (Nolan, J., dissenting); *id.* at 1275 (O'Connor, J., dissenting); *International Shoe Co. v. Washington,* 326 U.S. 310, 316 (1945); *Tauza v. Susquehanna Coal Co.,* 115 N.E. 915, 917 (N.Y. 1917); *Pramatha Nath Nullick v. Pradyumna Kumar Mullick,* 52 Indian L. R. 245, 250 (India 1925). *See* Roscoe Pound, *supra* note 37, at 195, 197, 198.

126. John Chipman Gray, *supra* note 39, at 43.

127. Jeremy Bentham, "The elements of the art of packing as applied to special juries" (Garland Publishing 1978) (1821) (emphasis in the original).

128. Tom Regan, *The Case for Animal Rights* 84–85 (Temple University Press 1983); James Rachels, *Created from Animals* 140, 147 (Oxford University Press 1990); William A. Wright, "Treating animals as ends," *J. Value Inquiry* 353, 357, 362 (1993); Christopher Cherniak, *Minimal Rationality* 3–17 (MIT Press 1985).

129. XII *Oxford English Dictionary* 269–270, definitions A.I.1.a and b; A.I.2.a, b, and c; A.I.3. A.I.4 (2d ed. 1989).

130. Eric Rakowski, *Equal Justice* 359 (Oxford University Press 1991).

131. Joel Feinberg, "Potentiality, development, and rights," in *The Problem of Abortion* 145 (2d ed., Joel Feinberg, ed., Wadsworth Publishing 1984) (emphases in the original). *See also* H. Tristram Englehardt, Jr., *The Foundations of Bioethics* 143 (2d ed., Oxford University Press 1996). *See* Stanley I. Benn, "Abortion, infanticide, and respect for persons," in Joel Feinberg, ed., *supra* at 143.

132. Isaiah Berlin, *supra* note 20, at 208.

133. Steven M. Wise, *supra* note 98, at 243–244.

134. Elinor Molbegott, review of *Rattling the Cage: Toward Legal Rights for Animals, New York Law Journal* at 2 (March 28, 2000).

135. Cass R. Sunstein, "The chimps' day in court," *New York Times Book Review* 26 (February 20, 2000) (emphasis added).

136. Betty Daror, "Bentham's law," *New York Times Book Review* 2 (March 19, 2000).

Chapter 3 Who Gets Liberty Rights?

1. Donald R. Griffin, *Animal Minds: Beyond Cognition to Consciousness* 11 (University of Chicago Press 2001).

2. *Id.* at 12.

3. Gordon G. Gallup, Jr., "Chimpanzees: Self-recognition," 167 *Science* 86 (1970).

4. Bernd Heinrich, "Testing insight in ravens," in *The Evolution of Cognition* 289, 289, 300–301 (Cecelia Heyes and Ludwig Huber, eds., MIT Press 2000); Sandra T. deBlois *et*

al., "Object permanence in orangutans *(Pongo pygmaeus)* and squirrel monkeys *(Saimiri sciureus),*" 112 *Journal of Comparative Psychology* 137, 148 (1998); Michael Tomasello and Josep Call, *Primate Cognition* 68–69 (Oxford University Press 1997).

5. Bernd Heinrich, *The Mind of the Raven: Investigations and Adventures with Wolf-Birds* 337 (Cliff Street Books 1999).

6. Antonio Damasio, *The Feeling of What Happens: Body and Emotion in the Making of Consciousness* 106 (Harcourt Brace 1999).

7. Donald R. Griffin, *supra* note 1, at 10.

8. Gerald M. Edelman and Giulio Tononi, *A Universe of Consciousness: How Matter Becomes Imagination* 107 (Basic Books 2000); Gerald M. Edelman, *Bright Air, Brilliant Fire: On the Matter of Mind* 123 (Basic Books 1992).

9. Michel Cabanac, "Emotion and phylogeny," 6 *Journal of Consciousness Studies* 176 (1999).

10. Donald R. Griffin, *supra* note 1, at 10.

11. Martha C. Nussbaum, *Poetic Justice: The Literary Imagination and Public Life* 38 (Beacon Press 1995).

12. Charmian Barton, "The status of the precautionary principle in Australia: Its emergence in legislation and as a common law doctrine, 12 *Harvard Environmental Law Review* 509, 510, 547 (1998); James E. Hickey, Jr., and Vern R. Walker, "Refining the precautionary principle in international environmental law," 14 *Virginia Environmental Law Journal* 423, 448 (1995).

13. W. Page Keeton *et al., Prosser and Keeton on the Law of Torts,* sec. 41, at 263, 269 (5th ed. 1984).

14. Naomi Roht-Arriaza, "Precaution, participation, and the 'greening' of international trade law," 7 *Journal of Environmental Law and Litigation* 57, 60–61 (1992); Ellen Hey, "The precautionary concept in environmental policy and law: Institutionalizing caution," 4 *Georgetown International Environmental Law Review* 303, 307–308 (1992).

15. Naomi Roht-Arriaza, *supra* note 14, at 60.

16. Ellen Hey, *supra* note 14, at 311; Charmian Barton, *supra* note 12, at 513, quoting Daniel Bodansky, "Scientific uncertainty and the precautionary principle," 4 *Environment* 4 (September 1991).

17. Michele Territo, "The precautionary principle in marine fisheries conservation and the U.S. Sustainable Fisheries Act of 1996," 24 *Vermont Law Review* 1351, 1352–1358 (1999); Charmian Barton, *supra* note 12, at 514–518; Chris W. Backes and Jonathan M. Verschuren, "The precautionary principle in International, European, and Dutch wildlife law," 9 *Colorado Journal of International Environmental Law and Policy* 43 (1998); David Bodansky, "The precautionary principle in U.S. environmental law," in *Interpreting the Precautionary Principle* 203–228 (Tim O'Riordan and James Cameron, eds., Cameron May, Ltd. 1996); James E. Hickey, Jr., and Vern R. Walker, *supra* note 12, at 432–436; Sonja Boehmer-Christiansen, "The precautionary principle in Germany: Enabling government," in *id.* at 31–60; Nigel Haigh, "The introduction of the precautionary principle in the U.K.," in *id.* at 229–251; Ronnie Harding, "The precautionary principle in Australia," in *id.* at 262–283; James Cameron, "The status of the precautionary principle in international law," in *id.* at 262–283.

18. *Roosevelt Campobello International Park v. Environmental Protection Agency,* 684 F. 2d 104, 1049 (D.C. Cir. 1982) (Endangered Species Act); *Committee for Humane Legisla-*

tion v. Richardson, 540 F. 2d 1141, 1145 (D.C. Cir. 1976) (Marine Mammal Protection Act).

19. World Charter for Nature, A/Res/37/7 art. 11/b (October 28, 1982).

20. Barnabas Dickson, "The precautionary principle in CITES: A critical assessment," 39 *Natural Resources Journal* 211, 213 (1999).

21. James E. Hickey, Jr., and Vern R. Walker, *supra* note 12, at 448–449.

22. Jonas Ebbeson, *Compatibility of International and National Environmental Law* 119 n. 73 (Kluwer Law International 1996).

23. Charmian Barton, *supra* note 12, at 520, 550; David Freestone, "The precautionary principle," in *International Law and Global Climate Change* 25 (Robin Churchill and David Freestone, eds., International Environmental Law and Policy Series 1991); J. Cameron and J. Abouchar, "The precautionary principle: A fundamental principal of law and policy for the protection of the global environment," 14 *Boston College International & Comparative Law Review* 1, 20–23 (1991). *See, e.g.*, David Pearce, "The precautionary principle and economic analysis," in *id.* at 132, 144.

24. William Rogers, "Benefits, costs and risks: Oversight of health and environmental decision-making," 4 *Harvard Environmental Law Review* 191, 225 (1980).

25. *American Bald Eagle v. Bhatti,* 9F. 3d 163 (1st. Cir. 1993).

26. Christina Hoff, "Immoral and moral uses of animals," 302 *New England Journal of Medicine* 115, 117 (1980).

27. Hugh Thomas, *The Slave Trade* 309 (Simon & Schuster 1997), quoting John Newton, "Thoughts on the African slave trade," in *Letters and Sermons* 103 (1780).

28. *E.g., Tumey v. Ohio,* 273 U.S. 510, 522 (1927).

29. *Bonham's Case,* 8 Co. 114a, 118a (C.P. 1610). *See* Harold J. Cook, "Against common rights and reason: The College of Physicians versus Dr. Thomas Bonham," 29 *The American Journal of Legal History* 301, 315–316 (1985); George P. Smith, "Dr. Bonham's case and the modern significance of Lord Coke's influence," 41 *Washington Law Review* 297, 301–304 (1966); Theodore F. T. Plucknett, "Bonham's case and judicial review," 40 *Harvard Law Review* 30, 32, 34 (1926).

30. *Tumey, supra* note 28, at 524, 525.

31. Fugitive Slave Act of 1850. The text can be read in William Lloyd Garrison's *Liberator* of September 27, 1850.

32. For a detailed argument that this belief is error, *see* Ray Greek and Jean Swingle Greek, *Sacred Cows and Golden Geese* (Continuum International Publishing Group 2000).

33. *See, e.g., United States v. Will,* 449 U.S. 200, 213–216 (1980).

34. Laurence H. Tribe, *American Constitutional Law* 1518 (2d ed., Foundation Press 1988), quoting *Strauder v. West Virginia,* 100 U.S. 303, 306 (1880).

35. R. H. Bradshaw advocates treating all animals alike, R. H. Bradshaw, "Consciousness in non-human animals: Adopting the precautionary principle," 5 *Journal of Consciousness Studies* 108 (1998).

36. Carl Wellman, *Real Rights* 129 (Oxford University Press 1995); Alan Gewirth, *Reason and Morality* 111, 121 (University of Chicago Press 1978).

37. Carl Wellman, *supra* note 36, at 130–131.

38. Henry Beston, *The Outermost House: A Year of Life on the Great Beach of Cape Cod* 25 (Penguin Books 1988).

39. Stuart Sutherland, *International Dictionary of Psychology* 211 (Continuum Publishing 1989).

40. Bernd Heinrich, *supra* note 5, at 327.

41. Diane Reiss, "The dolphin: An alien intelligence," in *First Contact: The Search for Extraterrestrial Intelligence* 31, 39 (Ben Bova and Bryon Press, eds., 1990).

42. Thomas Nagel, "What is it like to be a bat?" 83(4) *Philosophical Review* 435 (1974), reprinted in *The Nature of Consciousness: Philosophical Debates* 519, 520 (Ned Block *et al.*, eds., MIT Press 1997).

43. *Id.* (emphasis in the original). *See id.* at 520–521.

44. Frans de Waal, "Foreword," in *Anthropomorphism, Anecdotes, and Animals* xiv (State University of New York Press 1997) ("if closely related species act the same, the underlying mental processes are probably the same, too"); Frans M. B. de Waal, *Good-Natured: The Origins of Right and Wrong in Humans and Other Animals* 64–65 (Harvard University Press 1996); Frans M. B. de Waal, "The chimpanzee's sense of social regularity and its relation to the human sense of justice," 34 *American Behavioral Scientist* 335, 341 (Jan./Feb. 1991) ("strong arguments would have to be furnished before we would accept that similar behaviors in related species are differently motivated").

45. Georgia J. Mason *et al.*, "Frustrations of fur-farmed mink," 410 *Nature* 35, 35–36 (2001).

46. William R. Langbauer, Jr., *et al.*, "Responses of captive African elephants to playback of low frequency calls," 67 *Canadian Journal of Zoology* 2604 (1989).

47. Joyce Poole, *Coming of Age with Elephants* 114 (Hodder & Stoughton 1996).

48. Harry F. Harlow, "The evolution of learning," in *Behavior and Evolution* 269 (Anne Roe and Gaylord Simpson, eds., Yale University Press 1958).

49. Harry Jerison, "The perceptual worlds of dolphins," in *Dolphin Cognition and Behavior: A Comparative Approach* 137, 148, 149 (Ronald J. Schusterman *et al.*, eds., Lawrence Erlbaum 1986).

Chapter 4 Christopher

1. *See* Marc D. Hauser, *Wild Minds: What Animals Really Think* vxii (Henry Holt 2000).

2. *Id.* at 33–37, 154–169, 175.

3. Michael Cole and Sheila R. Cole, *The Development of Children* 33, 37 (Worth Publishers 2001).

4. Noam Chomsky, *Language and Problems of Knowledge* 161 (MIT Press 1988).

5. Elizabeth S. Spelke *et al.*, "Early knowledge of object motion: Continuity and inertia," 51 *Cognition* 131, (1994); Annette Karmiloff-Smith, *Beyond Modularity: A Developmental Perspective on Cognitive Science* (MIT Press 1992).

6. *See* Franck Ramus *et al.*, "Language discrimination by human newborns and by cotton-top tamarin monkeys," 288 *Science* 349 (April 14, 2000); S. J. Paterson *et al.*, "Cognitive modularity and genetic disorders," 286 *Science* 2355 (December 17, 1999); Dorothy V. M. Bishop, "An innate basis for language," 286 *Science* 2283 (December 17, 1999); Sue Taylor Parker and Michael L. McKinney, *Origins of Intelligence: The Evolution of Cognitive Development in Monkeys, Apes, and Humans* 189 (Johns Hopkins Press 1999).

7. Jean M. Mandler, "Representation," in 2 *Handbook of Child Psychology* 272, 277–280 (5th ed., William Damon, ed.-in-chief, Deanna Kuhn and Robert S. Siegler, vol. eds.,

John Wiley & Sons, Inc. 1998). *See* Alison Gopnik *et al., The Scientist in the Crib: Minds, Brains, and How Children Learn* 7, 16–17, 25, 120, 131, 142–143, 147–159 (William Morrow and Co. 1999); Rochelle Gelman, "Epigenetic foundations of knowledge structures: Initial and transcendent constructions," in *The Epigenesis of Mind: Essays on Biology and Cognition* 203, 312 (Susan Carey and Rochelle Gelman, eds., Lawrence Erlbaum 1991). For example, as to cause and effect, *see id.* at 282; David Premack, "The infant's theory of self-propelled objects," 36 *Cognition* 1 (1990).

8. Marshall M. Haith and Janette B. Benson, "Infant cognition," in William Damon, ed.-in-chief, Deanna Kuhn and Robert S. Siegler, vol. eds., *supra* note 7, at 244.

9. Elizabeth S. Spelke, "Initial knowledge: Six suggestions," 50 *Cognition* 431, 431 (1994).

10. Marshall M. Haith and Janette B. Benson, *supra* note 8, at 200, quoting H. Beilin, "Jean Piaget's enduring contribution to developmental psychology," in *A Century of Developmental Psychology* 257 (R. D. Parke, P. A. Ornstein, J. J. Reiser, and C. Zahn-Wexler, eds., American Psychological Association), reprinted from 28 *Developmental Psychology* 191.

11. Annette Karmiloff-Smith, *supra* note 5.

12. Personal communication from Jean M. Mandler, dated September 22, 2000.

13. Personal communication from Jean M. Mandler, dated October 29, 2000.

14. Henry M. Wellman, *The Child's Theory of Mind* 317 (MIT Press 1992).

15. Marshall M. Haith and Janette B. Benson, *supra* note 8, at 206.

16. *E.g.*, Michael Tomasello and Josep Call, *Primate Cognition* 26 (Oxford University Press 1997); Claude Dumas and Donald M. Wilkie, "Object permanence in ring doves *(Streptopelia risoria),*" 109 *Journal of Comparative Psychology* 142, 150 (1997).

17. Anne E. Russon, "Naturalistic Approaches to Orangutan Intelligence and the Question of Enculturation," 12 *International Journal of Comparative Psychology* 181, 184 (May, 2000), referring to Sue Taylor Parker and Michael L. McKinney, *supra* note 6, and Michael Tomasello and Josep Call, *supra* note 16.

18. Kim Bard, "Sensorimotor cognition in young feral orangutans," 36 *Primates* 297, 298 (1985).

19. Sue Taylor Parker, "Piaget's sensorimotor period series in an infant macaque: A model for comparing unstereotyped behavior and intelligence in human and nonhuman primates," in *Primate Biosocial Development* 43, 43 (S. Chevalier-Skolnikoff and F. Poirier, eds., Garland 1997).

20. Personal communication from Anne E. Russon, dated October 5, 2000.

21. Irene Pepperberg *et al.*, "Development of Piagetian object permanence in a grey parrot *(Psittacus erithacus),* 111 *Journal of Comparative Psychology* 63, 73 (1997).

22. Sue Taylor Parker and Michael L. McKinney, *supra* note 6, at 25–27; J. H. Flavell, *The Developmental Psychology of Jean Piaget* 86, note 1, 87–89 (D. V. Nostrand Co. 1963); Jean Piaget, *The Origins of Intelligence in Children* 20–356 (Margaret Cook, trans., W. W. Norton & Co. 1952).

23. Personal communication from Jean M. Mandler, dated September 22, 2000.

24. Robert Kegan, *The Evolving Self: Problem and Process in Human Development* 30 (Harvard University Press 1982).

25. Bernd Heinrich, "Testing insight in ravens," in *The Evolution of Cognition* 289, 289, 300–301 (Cecelia Heyes and Ludwig Huber, eds., MIT Press 2000); Sandra T. deBlois

et al., "Object permanence in orangutans *(Pongo pygmaeus)* and squirrel monkeys *(Saimiri sciureus)*," 112 *Journal of Comparative Psychology* 137, 148 (1998); Michael Tomasello and Josep Call, *supra* note 16, at 68–69.

26. J. H. Flavell, *supra* note 22, at 150.

27. Personal communication from Jean M. Mandler, dated September 22, 2000.

28. Robert Kegan, *supra* note 24, at 32; J. H. Flavell, *supra* note 24, at 156.

29. J. H. Flavell, *supra* note 22, at 157–159.

30. Sue Taylor Parker and Michael L. McKinney, *supra* note 6, at 28, 29.

31. J. H. Flavell, *supra* note 22, at 165.

32. *Id.* at 202–211. See Robert Kegan, *supra* note 22, at 38–39; Jerome Kegan, "A conception of early adolescence," in *Twelve to Sixteen: Early Adolescence* 90 (Jerome Kagan and Robert Coles, eds., W. W. Norton 1972).

33. Alison Gopnik *et al., supra* note 7, at 16, 73.

34. Robert Kegan, *supra* note 24, at 28.

35. Jean Piaget, *supra* note 22, at 357.

36. Jean M. Mandler, *supra* note 7, at 255, 266.

37. Jean Piaget, *The Psychology of Intelligence* 121 (Littlefield, Adams and Co., 1976); J. H. Flavell, *Cognitive Development* 44 (2d ed., Prentice Hall 1985); J. H. Flavell, *supra* note 22, at 152.

38. Jean M. Mandler, *supra* note 7, at 266, 300.

39. J. H. Flavell, *supra* note 22, at 151.

40. François Y. Doré and Sonia Goulet, "The comparative analysis of object knowledge," in *Piaget, Evolution, and Development* 55, 58, 62 (Jonas Langer and Melanie Killen, eds., Lawrence Erlbaum Associates 1998).

41. Personal communication from Jean M. Mandler, dated October 31, 2000.

42. Jean Piaget, *Play, Dreams and Imitation in Childhood* 238 (W. W. Norton 1962); Jean Piaget, *The Construction of Reality in the Child* 357, 358 (Margaret Cook, trans., Basic Books 1954).

43. J. H. Flavell, *supra* note 22, at 152.

44. Jean M. Mandler, *supra* note 7, at 265.

45. *Id.* at 266 (emphasis in the original).

46. *Id.* at 269–270, quoting Jean M. Mandler, "Representation," in 3 *Cognitive Development: Handbook of Child Psychology* 420, 424 (P. Mussen, series ed., J. H. Flavell and E. M. Markman, vol. eds., John Wiley and Sons, Inc. 1983).

47. Robert Kegan, *supra* note 24, at 30.

48. *Id.* at 31.

49. Jean M. Mandler, *supra* note 7, at 286. *See* Lorraine E. Bahrick, "Intermodal origins of self-perception," in *The Self in Infancy: Theories and Research* 349, 369 (Philippe Rochat, ed., Elsevier 1995); Peter D. Eimas and Paul C. Quinn, "Studies on the formation of perceptually based basic-level categories in young infants," 65 *Child Development* 903, 915 (1994).

50. Jean M. Mandler, "Perceptual and conceptual processes in infancy," 1 *Journal of Cognition and Development* 3, 3–4 (2000); Jean M. Mandler, *supra* note 7, at 272, 273–277, 286–293. *See* Michael Cole and Sheila R. Cole, *supra* note 3, at 202, 204.

51. Jean Piaget, *supra* note 22, at 23–46. *See* J. H. Flavell, *supra* note 37, at 21; J. H. Flavell, *supra* note 22, at 89–91.

52. Jean Piaget *(Play, Dreams and Imitation in Childhood), supra* note 42, at 8–18.

53. Jean Piaget, *supra* note 22, at 47–143. *See* Eugene Subbotsky, *Foundations of the Mind: Children's Understanding of Reality* 42 (Harvard University Press 1993); J. H. Flavell, *supra* note 37, at 21–22, 34; J. H. Flavell, *supra* note 22, at 91–101, 130; Robert Kegan, *supra* note 24 at 30.

54. J. H. Flavell, *supra* note 22, at 142–143.

55. Sue Taylor Parker and Michael L. McKinney, *supra* note 6, at 106; J. H. Flavell, *supra* note 22, at 124.

56. Alison Gopnik *et al., supra* note 7, at 74–75; J. H. Flavell, *supra* note 22, at 143–144.

57. Jean Piaget, *supra* note 22, at 171.

58. Sue Taylor Parker and Michael L. McKinney, *supra* note 6, at 57; J. H. Flavell, *supra* note 22, at 138–139.

59. Jean Piaget, *supra* note 37, at 21–22, 36–37. *See* Marshall M. Haith and Janette B. Benson, *supra* note 8, at 207–208; Eugene Subbotsky, *supra* note 53, at 42; J. H. Flavell, *supra* note 37, at 35; J. H. Flavell, *supra* note 22, at 131–132.

60. Jean Piaget *(Play, Dreams and Imitation in Childhood), supra* note 42, at 18–29; Jean Piaget, *supra* note 22, at 147–209. *See* Sue Taylor Parker and Michael L. McKinney, *supra* note 6, at 107; J. H. Flavell, *supra* note 37, at 22–24, 35; J. H. Flavell, *supra* note 22, at 101–109, 125, 130–132.

61. Michael Cole and Sheila R. Cole, *supra* note 3, at 189.

62. *Id.* at 191. *See* the summary of the stages of object permanence, Sandra T. de Blois *et al.,* "Object permanence in orangutans *(Pongo pygmaeus)* and squirrel monkeys *(Samiri sciureus),* 112 *Journal of Comparative Psychology* 137, 138 (1998).

63. Michael Cole and Sheila R. Cole, *supra* note 3, at 187–188, 191, 193.

64. Barbel Inhelder and Jean Piaget, *The Early Growth of Logic in the Child* 5, 13–15 (W. W. Norton & Co. 1969).

65. Jean Piaget, *supra* note 22, at 210–262. *See* Sue Taylor Parker and Michael L. McKinney, *supra* note 6, at 45; J. H. Flavell, *supra* note 37, at 24–25; J. H. Flavell, *supra* note 22, at 109–113, 145.

66. Sue Taylor Parker and Michael L. McKinney, *supra* note 6, at 75; J. H. Flavell, *supra* note 22, at 139–140.

67. Jean Piaget *(Play, Dreams and Imitation in Childhood), supra* note 42, at 30–52. *See* Sue Taylor Parker and Michael L. McKinney, *supra* note 6, at 107; J. H. Flavell, *supra* note 22, at 125.

68. Jean Piaget, *supra* note 37, at 123.

69. Michael Cole and Sheila R. Cole, *supra* note 3, at 191.

70. Lise Elliot, *What's Going On in There?: How the Brain and Mind Develop in the First Five Years of Life* 409 (Bantam Books 1999).

71. François Y. Doré and Sonia Goulet, "The comparative analysis of object knowledge," in Jonas Langer and Melanie Killen, eds., *supra* note 40, at 55, 59; Lawrence Erlbaum Associates 1998); Irene Pepperburg *et al., supra* note 21, at 111.

72. Jean Piaget *(The Construction of Reality in the Child), supra* note 42. *See* J. H. Flavell, *supra* note 37, at 36; J. H. Flavell, *supra* note 22, at 132–133.

73. *E.g.,* Lise Elliot, *supra* note 70, at 409; Irene Maxine Pepperberg, *The Alex Studies: Cognitive and Communicative Abilities of Grey Parrots* 190–191 (Harvard University Press

1999); Jean M. Mandler, *supra* note 7, at 272, 297; Marshall M. Haith and Janette B. Benson, *supra* note 8, at 207–208; Irene Pepperberg *et al., supra* note 21, at 71.

74. J. H. Flavell, *supra* note 22, at 145–146.

75. J. H. Flavell, *supra* note 37, at 36–37; J. H. Flavell, *supra* note 22, at 133–134.

76. Jean Piaget, *supra* note 22, at 263–330. *See* J. H. Flavell, *supra* note 37, at 24–25; J. H. Flavell, *supra* note 22, at 113–118.

77. J. H. Flavell, *supra* note 22, at 140–141.

78. Sue Taylor Parker and Michael L. McKinney, *supra* note 6, at 107; J. H. Flavell, *supra* note 22, at 125–126; Jean Piaget *(Play, Dreams and Imitation in Childhood), supra* note 42, at 52–61.

79. Sue Taylor Parker and Michael L. McKinney, *supra* note 6, at 45.

80. J. H. Flavell, *supra* note 22, at 145–146.

81. J. H. Flavell, *supra* note 37, at 36–37; J. H. Flavell, *supra* note 22, at 133–134.

82. Michael Tomasello and Josep Call, *supra* note 16, at 42; J. H. Flavell, *supra* note 22, at 151; Jean Piaget, *supra* note 22, at 355.

83. J. H. Flavell, *supra* note 37, at 37; J. H. Flavell, *supra* note 22, at 134, 141–142.

84. J. H. Flavell, *supra* note 22, at 136.

85. Jean Piaget *(Play, Dreams and Imitation in Childhood), supra* note 42, at 62–86; Jean Piaget, *supra* note 22, at 331–356. *See* Sue Taylor Parker and Michael L. McKinney, *supra* note 6, at 28, 107; J. H. Flavell, *supra* note 37, at 26–33; J. H. Flavell, *supra* note 22, at 126, 152.

86. J. H. Flavell, *supra* note 22, at 151; Jean Piaget *(Play, Dreams and Imitation in Childhood), supra* note 42, at 238.

87. *See, e.g.,* Andrew Whiten, "Chimpanzee cognition and the question of mental re-representation," in Daniel Sperber, *Metarepresentation: A Multidisciplinary Perspective* 1, 3 (Oxford University Press 2000); Josef Perner, *Understanding the Representational Mind* 16, 280 (MIT Press 1991).

88. Jean Piaget, *supra* note 37, at 120–121; Jean Piaget *(Play, Dreams and Imitation in Childhood), supra* note 42, at 67–68, 273.

89. Michael Cole and Sheila R. Cole, *supra* note 3, at 224.

90. Marc D. Hauser, *supra* note 1, at 26–26; Jean M. Mandler, *supra* note 7, at 271.

91. Rene Baillargeon, "Representing the existence and the location of hidden objects: Object permanence in 6- and 8-month old infants," 23 *Cognition* 21 (1986).

92. Jean M. Mandler, *supra* note 7, at 271. *See id.* at 280–281.

93. Henry A. Wellman and Susan A. Gelman, "Knowledge acquisition in foundational domains," in William Damon, ed.-in-chief, Deanna Kuhn and Robert S. Siegler, vol. eds., *supra* note 7, at 523, 530, 531.

94. Jean M. Mandler, *supra* note 7, at 269.

95. Personal communication from Jean M. Mandler, dated October 29, 2000.

96. Jean M. Mandler, *supra* note 7, at 299–300.

97. Alison Gopnik *et al., supra* note 7, at 27.

98. Jean Mandler, *supra* note 7 at 271–272 citing L. McDonough, "Memory for location in 7-month-olds," (1997) (manuscript submitted for publication).

99. Elizabeth Hannah and Andrew N. Meltzoff, "Peer imitation by toddlers in laboratory, home, and daycare contexts: Implications for social learning and memory," 29(4) *Developmental Psychology* 701 (1993).

100. Marshall M. Haith and Janette B. Benson, *supra* note 8, at 235–241.

101. Michael Tomasello and Josep Call, *supra* note 16, at 42

102. Marshall M. Haith and Janette B. Benson, *supra* note 8, at 205.

103. *Id.* at 207 (emphases in original).

104. Jean M. Mandler, *supra* note 7, at 270; Marshall M. Haith and Janette B. Benson, *supra* note 8, at 205, 202. *See* K. W. Fischer and T. Bidell, "Constraining nativist inferences about cognitive capacities," in Susan Carey and Rochelle Gelman, eds., *The Epigenesis of Mind: Essays on Biology and Cognition, supra* note 7.

105. Michael Cole and Sheila R. Cole, *supra* note 3, at 197–198.

106. *Id.* at 198.

107. Michael Cole and Sheila R. Cole, *supra* note 3, at 344; Jean M. Mandler, *supra* note 7, at 269; Marshall M. Haith and Janette B. Benson, *supra* note 8, at 211 (objects), 214 (space), 218 (time), 244; Michael Tomasello and Josep Call, *supra* note 16 at 38, 172.

108. Michael Tomasello and Josep Call, *supra* note 16, at 318, 405.

109. Colin Allen and Marc Bekoff, *Species of Mind: The Philosophy and Biology of Cognitive Ethology* 14 (MIT Press 1997); Daniel C. Dennett, *Kinds of Minds* 36 (Basic Books 1996); Daniel C. Dennett, *The Intentional Stance* 242–243 (MIT Press 1987).

110. Daniel Dennett *(Kinds of Minds), supra* note 109, at 34.

111. Susan Greenfield, *The Private Life of the Brain* 170, 227, note 16 (The Penguin Press 2000).

112. The affective neuroscientist, Jaak Panksepp, notes that "formal analysis of rat behavior has led to the conclusion that such creatures do exhibit some true intentionality," Jaak Panksepp, *Affective Neuroscience: The Foundations of Human and Animal Emotions* 306 (Oxford University Press 1998).

113. John Searle, *The Rediscovery of the Mind* 78–79 (MIT Press 1992).

114. Daniel C. Dennett *(The Intentional Stance), supra* note 109, at 271.

115. Martha C. Nussbaum, *Upheavals of Thought: The Intelligence of Emotions* 27–28, 91–126 (Cambridge University Press 2001).

116. Personal communication from Antonio R. Damasio, dated September 6, 2000; Antonio Damasio, *The Feeling of What Happens: Body and Emotion in the Making of the Consciousness* 116–117 (Harcourt Brace 1999) (emphasis added).

117. Daniel C. Dennett *(The Intentional Stance), supra* note 109, at 243. Child psychologist Josef Perner uses his own terminology: He argues that Christopher will pass three mental stages on his way to his complete understanding of minds. Stage One was "primary representation," Josef Perner, *supra* note 87 at 6. Here Christopher could represent the outside world inside his mind more or less as it was, Andrew Whiten, "Imitation, pretense, and mindreading," in *Reaching Into Thought: The Minds of the Great Apes* 300, 313 (Anne E. Russon *et al.,* eds., Cambridge University Press 1996). In his second year, Christopher attained what child psychologist Alan M. Leslie calls "metarepresentation," Alan M. Leslie, "Pretense and representation in infancy: The origins of 'theory of mind,'" 94 *Psychological Review* 412, 414–416 (1987). The primatologist Andrew Whiten refers to this as "re-representation," Andrew Whiten, *supra* note 87, at 2–4. In its most complex form, Perner calls it "secondary representation," Josef Perner, *supra* note 87, at 413. This is the ability to imagine hypothetical situations, *Id.* at 2–4; Andrew Whiten, *supra (Reaching Into Thought),* at 312–314; Josef Perner, *supra* note 87, at 7, Alan M. Leslie, *supra* at 413. Whatever you call it, Christopher began to attribute mental

states to others. Sometime during his fifth year, he will attain what Perner calls "metarepresentation" (Warning! This is different from Alan Leslie's "metarepresentation"). Now he will *understand* that minds can represent and that other minds may contain different ideas from his own, Sue Taylor Parker and Michael L. McKinney, *supra* note 6, at 137; Andrew Whiten, *supra (Reaching Into Thought),* at 312–314; Josef Perner, *supra* note 87, at 7; Alan M. Leslie, *supra* at 421–422.Although Piaget, Perner, Dennett, Whiten, and Leslie may not agree on many of the fine points, they are all talking about similar things. Thus Dennett's zero order of intentionality roughly corresponds to Piaget's Stages 1 and 2. His first order of intentionality is about the same as Piaget's Stages 3 through 5. Stages 1 through 5 approximate Perner's primary representation. Dennett's second order of intentionality roughly equals Perner's secondary representation as well as Piaget's sensorimotor Stage 6 and the symbolic subperiod of Piaget's peroperations period. Dennett's third (and higher) order equals Perner's metarepresentation and the intuitive subperiod of Piaget's preoperations period, Sue Taylor Parker and Michael L. McKinney, *supra* note 6, at 135–147; Andrew Whiten and Josef Perner, "Fundamental issues in the multidisciplinary study of mindreading," in *Natural Theories of Mind: Evolution, Development and Simulation of Everyday Mindreading* 1, 16–17 (Andrew Whiten, ed., Basil Blackwell 1991).

118. Alison Gopnik *et al., supra* note 7.

119. *E.g.*, Marc Hauser and Susan Carey, "Building a cognitive creature from a set of primitives," in *The Evolution of Mind* 51, 58–59 (Denise Dellarose Cummins and Colin Allen, eds., Oxford University Press 1998); Elizabeth S. Spelke *et al.*, "Infants knowledge of object motion and human action," *Causal Cognition, A Multi-Disciplinary Debate* 44, in Dan Sperber *et al.,* eds., Oxford University Press 1995); Rene Baillargeon, "Reasoning about the height and location of a hidden object in 4.5 and 6.5 month old infants," 38 *Cognition* 13, 37 (1991).

120. Andrew Whiten and Josef Perner, *supra* note 117, at 13.

121. *E.g.,* Henry M. Wellman, "From desires to beliefs: Acquisition of a theory of mind," in *Natural Theories of Mind: Evolution, Development and Simulation of Everyday Mindreading* 19, 28–32 (Andrew Whiten, ed., Basil Blackwell 1991); Henry M. Wellman, *supra* note 14, at 210–213; Tom Regan, "Frey on why animals cannot have simple desires," 91 *Mind* 277 (1982); Tom Regan, "Feinberg on what sorts of beings can have rights," *Southern Journal of Philosophy* 485, 487 (Winter 1976).

122. *E.g.,* R.G. Frey, *Interests and Rights: The Case Against Animals* 28–37, 87–91, 169 (Oxford University Press 1980); Joel Feinberg, "The rights of animal and unborn generations," in *Philosophy and Environmental Crisis* 43 (W. T. Blackstone, University of Georgia Press 1974).

123. *E.g.*, R. G. Frey, *supra* note 122, at 28–37, 87–91, 169. *See* Antonio Damasio, *supra* note 116, at 107.

124. Antonio Damasio, *supra* note 116, 155; Marshall M. Haith and Janette B. Benson, *supra* note 8, at 200; Janet Wilde Astington, *The Child's Discovery of the Mind* 42 (Harvard University Press 1993); Jean Piaget *(Play, Dreams and Imitation in Childhood), supra* note 42, at 68.

125. Mark A. Schmuckler, "Self-knowledge of body position: Integration of perceptual and action system information," in Philippe Rochat, ed., *supra* note 49, at 221, 235; Andrew N. Meltzoff and Keith Moore, "A theory of the role of imitation in the emergence

of self," in Philippe Rochat, ed., *supra* note 49, at 73, 89; Philippe Rochat, "Early objectification of the self," in Philippe Rochat, ed., *supra* note 49, at 60, 63; Ulrich Neisser, "Criteria for an ecological self," in Philippe Rochat, ed., *supra* note 49, at 23.

126. J. H. Flavell, *supra* note 22, at 107–108; Jean Piaget, *supra* note 22, at 147–152; Jean Piaget, *supra* note 37, at 101.

127. Marshall M. Haith and Janette B. Benson, *supra* note 8, at 200, 207; Jean M. Mandler, *supra* note 7, at 270–271.

128. Emanuela Cenami Spada *et al.*, "The self as reference point: Can animals do without it?" in Philippe Rochat, ed., *supra* note 49, at 193, 195–196.

129. John S. Watson, "Self-orientation in early infancy: The general role of contingency and the specific case of reaching to the mouth," in Philippe Rochat, ed., *supra* note 49, at 375, 382.

130. *See* Gordon G. Gallup, Jr., "Chimpanzees: Self-recognition," 167 *Science* 86 (1970).

131. *E.g.*, Marc D. Hauser, *supra* note 1 at 101.

132. *Compare* György Gergely, "From self-recognition to theory of mind," in Sue Taylor Parker *et al.*, eds., *Self-Awareness in Animal and Humans* 51 (Cambridge University Press 1994) (does) *with* Maria L. Boccia, "Mirror behavior in Macaques," in *id.* at 350 (doesn't).

133. *Compare* Alison Gopnik and Andrew W. Meltzoff, "Minds, bodies, and persons: Young children's understanding of the self and others as reflected in imitation and theory of mind research," in Sue Taylor Parker *et al.*, eds., *id.* at 133 (does); Deborah Custance and Kim A. Bard, "The comparative and developmental study of self-recognition and imitation: The importance of social factors," in *id.* at 207 (doesn't).

134. Gordon G. Gallup, Jr., "Self-recognition: Research strategies and experimental design," in *id.* at 35 (does) with Alison Gopnik and Andrew W. Meltzoff, *supra* note 133, at 166; György Gergely, *supra* note 132 at 51 (doesn't).

135. Sue Taylor Parker and Michael L. McKinney, *supra* note 6, at 153, 157, 192; Sue Taylor Parker, "The evolution and development of self-knowledge: Adaptations for assessing the nature of self relative to that of conspecifics," in Jonas Langer and Melanie Killen, eds., *supra* note 40, at 171, 179.

136. Rene Zarro, "The person: Objective approaches," 6 *Review of Child Development Research* 247, 276–279 (W. W. Hartrup, ed., University of Chicago Press 1982).

137. Robert W. Mitchell, "Mental models of mirror self-recognition: Two theories," 11 *New Ideas in Psychology* 295, 295–312 (1993).

138. Louis J. Moses, "Foreword," in Sue Taylor Parker, *et al.*, eds., *supra* note 132, at 131, 138–144.

139. *E.g.*, Lorraine E. Bahrick, "Intermodal origins of self-perception," in Philippe Rochat, ed., *supra* note 49, at 349, 357; Bennett I. Bertenthal and James L. Rose, "Two modes of perceiving the self," in *id.* at 302, 310; Pierre Mounoud, "From direct to reflexive (self-) knowledge: A recursive model," in *id.* at 141, 141; Alan Fogel, "Relational narratives of the prelinguistic self," in *id.* at 117, 117; Philippe Rochat, "Early objectification of the self," in Philippe Rochat, ed., *supra* note 49, at 53, 67; George Butterworth, "The self as an object of consciousness in infancy," in *id.* at 35, 47; Ulrich Neisser, *supra* note 125, at 17, 30; H. Lyn White Miles, "ME CHANTEK: The development of self-awareness in a signing orangutan," in Sue Taylor Parker *et al.*, eds., *supra* note 132, at 254, 258–268.

140. Sue Taylor Parker, "The evolution and development of self-knowledge: Adaptations for assessing the nature of self relative to that of conspecifics," in Jonas Langer and Melanie Killen, eds., *supra* note 40, at 171, 178–179; Francine G. P. Patterson and Ronald H. Cohn, "Self-recognition and self-awareness in lowland gorillas," in Sue Taylor Parker *et al., supra* note 132, at 273, 288; Karyl B. Swartz and Sian Evans, "Social and cognitive factors in chimpanzee and gorilla mirror behavior and self-recognition," in *id.* at 189, 203.

141. *Id.*

142. Marc D. Hauser, *supra* note 1, at 109.

143. Ulrich Neisser, *supra* note 125 at 30.

144. Marc D. Hauser, *supra* note 1, at 107–109; Marc D. Hauser *et al.*, "Self-recognition in primates: Phylogeny and the salience of species-typical traits," 92 *Proceedings of the National Academy of Sciences* 10811 (1995).

145. Robbie Case, "Stages in the development of the young child's first sense of self," in 11 *Developmental Review* 210, 210 (1991); William James, *Psychology* 176-195 (Fine Editions Press 1998) (1892).

146. Louis J. Moses, *supra* note 138, at xii.

147. Irene M. Pepperberg, "Object permanence in four species of psittacine birds: An African Grey parrot *(Psittacus erithacus),* an Illiger mini macaw *(Ara maracana),* a parakeet *(Melopsittacus undulatus),* and a cockatiel *(Nymphicus hollandicus),* 18 *Animal Learning and Behavior* 97, 98 (1990).

148. Case assumed, more than Piaget, that the "I self" depended both upon both cognition (thinking) and affect (feeling).

149. Robbie Case, *supra* note 145, at 216.

150. Sue Taylor Parker and Michael L. McKinney, *supra* note 6, at 151–152; John S. Watson, "Detection of self: The perfect algorithm," in Sue Taylor Parker *et al.*, ed., note 132, at 131, 138–144.

151. Robbie Case, *supra* note 145, at 218. *See id.* at 219.

152. Susan Greenfield, *supra* note 111, at 33, 37, 45, 47, 66, 143, 144, 149, 154, 158, 167.

153. *E.g., The Smile of a Dolphin: Remarkable Accounts of Animal Emotions* (Marc Bekoff, ed., Crown/Discovery Books 2000); *Cognitive Neuroscience of Emotion* (Richard D. Cane *et al.,* eds., Oxford University Press 2000); Jaak Panksepp, *supra* note 112, at 306. For a popular account of emotions in nonhuman animals, *see generally,* Jeffrey Moussaieff Masson and Susan McCarthy, *When Elephants Weep: the Emotional Lives of Animals* (Delacorte Press 1995).

154. Susan Greenfield, *supra* note 111, at 2, 16, 20, 21, 22, 41, 50, 78, 146, 163, 174, 175. Greenfield proposes, however, that emotions sweep away self-consciousness (which she calls "Self") and that the more that one is under the influence of emotions, the less one will be self-consciousness, *id.* at 21, 139, 185–186.

155. Jaak Panksepp, *supra* note 112, at 300, 310.

156. *Id.* at 308, 311. Panksepp frequently, but not always, uses "animals" and "mammals" as synonyms.

157. *Id.* at 309.

158. Martha C. Nussbaum, *supra* note 115, at 90. *See* her discussion of emotions in nonhuman animals at *id.* at 89–138.

159. *Id.* at 100–106, discussing L. Y. Abramson *et al.*, "Learned helplessness in humans: Critique and reformulations," 87 *Journal of Abnormal Psychology* 4874 (1978); Martin E. P. Seligman, *Helplessness: On Depression, Development, and Death* (W. H. Freeman 1975).

160. George Pitcher, *The Dogs Who Came to Stay* (Dutton 1995). *See also* Barbara Smuts, untitled essay, in *The Lives of Animals* 107 (Amy Gutmann, ed., Princeton University Press 1999).

161. Martha C. Nussbaum, *supra* note 115, at 124.

162. Jaak Panksepp, *supra* note 112, at 7.

163. Marian Stamp Dawkins, *Through Our Eyes Only: The Search for Animal Consciousness* 143 (W. H. Freeman/Spektrum 1993).

164. Bruce G. Charlton, Review of *The Feeling of What Happens*, in 93 *Journal of the Royal Society of Medicine* 99 (February 2000).

165. Raymond J. Dolan, "Feeling the neurobiological self," 401 *Nature* 847, 847 (October 29, 1999). *See* Antonio Damasio, *supra* note 116.

166. Bruce G. Charlton, *supra* note 164, at 99.

167. Thomas Metzinger, "The hint half-guessed," 281(5) *Scientific American* 125 (1999); William H. Calvin, "Coming to," *New York Times Book Review* 8 (October 24, 1999).

168. *Id.*

169. Steven M. Wise, *Rattling the Cage: Toward Legal Rights for Animals* 130 (Perseus Publishing 2000).

170. Antonio Damasio, *supra* note 116, at 154.

171. Antonio Damasio, "How the brain creates the mind," *Scientific American* 112, 115 (December, 1999).

172. Antonio Damasio, *supra* note 116, at 189, 319.

173. Edelman and Tononi agree that a protoself exists, Gerald M. Edelman and Giulio Tononi, *A Universe of Consciousness: How Matter Becomes Imagination* 174 (Basic Books 2000).

174. Alexander Star, "What feelings are like," *New York Times Magazine* 31 (May 7, 2000).

175. Antonio Damasio, *supra* note 116, at 25–26, 35–81. *See* Alexander Star, *supra* note 174, at 31.

176. Antonio Damasio, *supra note* 116, at 9, 323.

177. *Id.* at 16. *See id.* at 16, 17, 18, 22, 25, 26, 30, 106, 117–125, 170, 171–172, 174, 175, 183–184, 192–193. Gerald Edelman and Giulio Tononi roughly agree, Gerald M. Edelman and Giulio Tononi, *supra* note 173, at 174.

178. Antonio Damasio, *supra* note 116, at 108–122, 169, 171, 185–188, 196.

179. *Id.* at 89–91, 105–106, 163.

180. *Id.* at 125, 127.

181. *Id.* at 175, 176, 185.

182. *Id.* at 172.

183. *Id.* at 203.

184. *Id.* at 16, 17, 121, 172–173 174, 175, 197, 198.

185. *Id.* at 196, 201. Susan Greenfield has similar thoughts. Consciousness blooms into self-consciousness as experience deepens and refigures a brain's neural connections,

Susan A. Greenfield, *supra* note 111, at 13, 186. Note Damasio's emphasis that extended consciousness is not intelligence; it allows an animal to become aware of what there is to know, but intelligence permits her to manipulate what she knows in novel and appropriate responses, Antonio Damasio, *supra* note 116, at 198–199.

186. Antonio Damasio, *supra* note 116, at 108, 174, 198. Edelman and Tononi concur, Gerald M. Edelman and Giulio Tononi, *supra* note 173, at 197.

187. Greenfield thinks that human consciousness awakens in the womb, Susan Greenfield, *supra* note 111, at 165.

188. Antonio Damasio, *supra* note 116, at 175.

189. *Id.* at 202.

190. Ulrich Neisser, "The self perceived," in *The Perceived Self: Ecological and Interpersonal Sources of Self-Knowledge* 3 (Ulrich Neisser, ed., Cambridge University Press 1993); Ulrich Neisser, "Two perceptually given aspects of the self and their development," 11 *The Developmental Review* 197, 197 (1991).

191. Ann E. Bigelow, "The effect of blindness on the early development on the self," in Philippe Rochat, ed., *supra* note 49, at 327, 341.

192. Michael Tomasello, "Understanding the self as social agent," Eleanor J. Gibson, "Are we automata?" in Philippe Rochat, *supra* note 49, at 5; J. J. Gibson, *The Ecological Approach to Visual Perception* 126 (Houghton Mifflin 1979).

193. Ulrich Neisser, *supra* note 125, at 17, 18, 30; Eleanor J. Gibson, *supra* note 192, at 5. *See* Jean Piaget *(The Construction of Reality in the Child),* 4–13 (Margaret Cook, trans., Basic Books 1954).

194. Ulrich Neisser, *supra* note 125, at 28.

195. *Id.* at 31.

196. Philippe Rochat, *supra* note 125, at 57.

197. George Butterworth, *supra* note 139, at 35, 40.

198. Ulrich Neisser, *supra* note 125, at 31.

199. Philippe Rochat, *supra* note 125, at 57–58, 63; Ulrich Neisser, *supra* note 125, at 28, 31.

200. Ulrich Neisser, *supra* note 125, at 27.

201. Philippe Rochat, *supra* note 125, at 53. Rochat included all insects, too. I don't think the evidence supports that claim.

202. Andrew N. Meltzoff and Keith Moore, *supra* note 125, at 77; Philippe Rochat, *supra* note 125, at 53, 66; George Butterworth, *supra* note 139, at 43; Ulrich Neisser, *supra* note 127, at 18, 20, 21, 32.

203. George Butterworth, *supra* note 139, at 37; Ulrich Neisser, supra note 125, at 18, 20.

204. Ulrich Neisser, *supra* note 125, at 18–19.

205. *Id.* at 19.

206. *Id.* at 25.

207. *Id.* at 25–26.

Chapter 5 Honeybees

1. Donald R. Griffin, *Animal Minds: Beyond Cognition to Consciousness* 268 (University of Chicago Press 2001).

2. Personal communication from "Mr. Darwin," dated June 14, 2001.

3. Yohannes Haile-Selassie, "Late Miocene hominids from the Middle Awash, Ethiopia," 412 *Nature* 178 (2001); Mary Ellen Ruvulo, "Molecular phylogeny of the Hominids," 14 *Molecular and Biological Evolution* 248 (1996); David Pilbeam, "Genetic and morphological records of the Hominoid and hominid origins: A synthesis," 5 *Molecular Phylogenetics and Evolution* 155, 164 (1996).

4. Two personal communications from Cam Muir, dated November 7, 2000.

5. Steven M. Wise, *Rattling the Cage: Toward Legal Rights for Animals* 133 (Perseus Publishing 2000).

6. There may be significant differences in how genes are expressed and regulated in the chimpanzee and human brains, Dennis Normile, "Gene expression differs in human and chimp brains," 292 *Science* 44 (April 6, 2001).

7. R. E. Snodgrass, *Anatomy of the Honeybees* 256 (Comstock Publishing 1956).

8. Randolf Menzel and Martin Girufa, "Cognitive architecture of a mini-brain: The honeybee," 59(2) *Trends in Cognitive Sciences* 62 (2001); G. Bicker *et al.*, "Chemical neuroanatomy of the honey bee brain," *Neurobiology and Behavior of Honeybees* 202, 203 (Springer-Verlag 1987).

9. Randolf Menzel and Martin Girufa, *supra* note 8, at 63–70; G. Bicker *et al., supra* note 8, at 205.

10. Martin Girufa *et al.,* "The concepts of 'sameness' and 'difference' in an insect," 410 *Nature* 930, 932 (April 19, 2001).

11. M. Lehrer, "Why do bees turn and look?" 172 *Journal of Comparative Physiology A* 549, 549 (1993); M. E. Bitterman, "Vertebrate-invertebrate comparisons," in *Intelligence and Evolutionary Biology* 251 (H. J. Jerison and I. Jerison, eds., Springer 1988).

12. Randolf Menzel, "Learning traces in honeybees," in *Neurobiology and Behavior of Honeybees* 310, 310 (Randolf Menzel and Alison Mercer, eds., Springer-Verlag 1987).

13. M. E. Bitterman, "Comparative analysis of learning in honeybees," 24(2) *Animal Learning & Behavior* 1213, 137 (1996).

14. Louis M. Herman, "The language of animal language research: Reply to Schusterman and Geisiner," 38 *The Psychological Record* 349, 366 (1988).

15. Randolf Menzel and Martin Girufa, *supra* note 8, at 69.

16. *Id.* As to the existence of short-term, or working, memory, *compare* M. F. Brown and G. E. Demas, "Evidence for spatial working memory in honeybees *(Apia mellifera),*" 108 *Journal of Comparative Psychology* 344 (1994) (exists) with S. Burnmeister *et al.*, "Performance of honeybees in analogies of the rodent radial maze," 23(4) *Animal Learning and Behavior* 369 (1995) (existence questioned).

17. Brian H. Smith and Susan Cobey, "The olfactory memory of the honeybee Apis mellifera. II. Blocking between odorants in binary mixtures," 195 *Journal of Experimental Biology* 91, 104 (1994).

18. Han de Vries and Jacobus C. Biesmeijer, "Modelling collective foraging by means of individual behavior rules in honeybees," 44 *Behavior and Ecological Sociobiology* 109, 110 (1998).

19. M. V. Srinivasan *et al.,* "Honeybees link sights to smells," 396 *Nature* 637 (1998).

20. James L. Gould and Carol Grant Gould, *The Animal Mind* 106–110 (Scientific American 1994). *See* Randolf Menzel and Martin Girufa, *supra* note 8, at 68–69; Marc D. Hauser, *Wild Minds: What Animals Really Think* 154 (Henry Holt and Co. 2000).

21. Marc D. Hauser, *supra* note 20, at 154.

22. Carl Anderson *et al.,* "Worker allocation in insect societies: Coordination of nectar foragers and nectar receivers in honey bee *(Apis mellifera)* colonies," 46 *Behavioral Ecology and Sociobiology* 73, 73 (1999).

23. Thomas D. Seeley, "Thoughts on information and integration in honey bee colonies," 29 *Apidologie* 67 (1998).

24. Donald R. Griffin, *supra* note 1, at 196.

25. Fed C. Dyer, "Spatial memory and navigation by honeybees on the scale of the foraging range," 199 *Journal of Experimental Biology* 147, 153 (1996).

26. Donald R. Griffin, *supra* note 1, at 194.

27. W. Kirchner *et al.,* "Honeybee dance communication: Acoustical indication of direction in round dances," 75 *Naturwissenschaften* 629, 629–630 (1988).

28. Donald R. Griffin, *supra* note 1, at 192, 195–196.

29. Marc D. Hauser, *The Evolution of Communication* 498 (MIT Press 1996).

30. Mandyan V. Srinivasan *et al.,* "Honey bee navigation: Nature and calibration of the 'odometer,'" 287 *Science* 851, 851–853 (2000); Harold E. Esch and John E. Burns, "Distance estimation by foraging honeybees," 199 *Journal of Experimental Biology* 155, 160–162 (1996).

31. Harold E. Esch *et al.,* "Honeybee dances communicate distances by optic flow," 411 *Nature* 581, 582–583 (2001).

32. Marc D. Hauser, *supra* note 29, at 498–500.

33. Donald R. Griffin, *supra* note 1, at 196.

34. *Id.* at 196–197; William F. Towne and James L. Gould, "The spatial precision of the honeybees: Dance communication," 1 *Journal of Insect Behavior* 129 (1988).

35. Donald R. Griffin, *supra* note 1, at 198; T. D. Seeley and S. C. Buhrman, "Group decision making in swarms of honeybees," 45 *Behavioral Ecology and Sociobiology* 19, (1999).

36. Carl Anderson *et al., supra* note 22, at 76.

37. Thomas D. Seeley *et al.,* "The shaking signal of the honey bee informs workers to prepare for greater activity," 104 *Ethology* 10, 12 (1998).

38. Marc D. Hauser, *supra* note 29, at 501.

39. Thomas D. Seeley, "The tremble dance of the honey bee: Message and meanings," 31 *Behavioral Ecology and Sociobiology* 375, 375 (1992).

40. *Id.* at 382.

41. *Id.*

42. James C. Nieh, "The honey bee shaking signal: Function and design of a modulatory communication signal," 42 *Behavioral Ecology and Sociobiology* 23 (1998).

43. James C. Nieh, "The stop signal of honey bees: Reconsidering its message," 33 *Behavioral Ecology and Sociobiology* 51 (1993).

44. Donald R. Griffin, *supra* note 1, at 203.

45. *Id.* at 194, 203.

46. *Id.* at 205.

47. T. D. Seeley and S. C. Buhrman, "Group decision making in swarms of honey bees," 45 *Behavioral Ecology and Sociobiology* 19 (1999).

48. Donald R. Griffin, *supra* note 1, at 207, 209.

49. *Id.* at 268.

50. Personal communication from M. E. Bitterman, dated August 21, 2001; C. I. Abramson, "Aversive conditioning in honeybees," 100 *Journal of Comparative Psychology* 108 (1986).

51. Antonio Damasio, *The Feeling of What Happens* 198, 201 (Harcourt Brace 1999); William H. Calvin, "Coming to," *New York Times Book Review* 8 (October 24, 1999).

52. Antonio Damasio, *supra* note 51, at 16, 37, 69–70, 100, 101, 106, 107, 188, 195, 198, 201, 239, 311, 314.

53. *Id.* at 201.

54. Marc Bekoff, "Animal emotions: Exploring passionate natures," 50 *BioScience* 861, 861 (October, 2000).

55. Jaak Panksepp, "Emotional Development yields lots of 'stuff'. . . especially mind 'stuff' that emerges from brain 'stuff,'" in *The Nature of Emotions: Fundamental Questions* 399 (Paul Ekman and Richard J. Davidson, eds., Oxford University Press 1994).

56. Susan Greenfield, *The Private Life of the Brain* 150 (Oxford University Press 2000).

57. Marc D. Hauser, *supra* note 20, at 10, 14.

58. Jaak Panksepp, *Affective Neuroscience: The Foundations of Human and Animal Emotions* 3, 4, 51, 300, 306 (Oxford University Press 1998); Jaak Panksepp, "The periconscious substrates of consciousness: Affective states and the evolutionary origins of the self," 5 *Journal of Consciousness Studies* 566, 567, 580 (1998). *See* Douglas F. Watt, "Consciousness and emotion," review of Jaak Panksepp's *Affective Neuroscience: The Foundations of Human and Animal Emotions,* 6 *Journal of Consciousness Studies* 191 (1999).

59. Edelman and Tononi say that language is not necessary for secondary consciousness, but is for the "me self" to emerge, Gerald M. Edelman and Giulio Tononi, *A Universe of Consciousness: How Matter Becomes Imagination* 103, 104, 174, 193, 197, 206 (Basic Books 2000); Gerald M. Edelman, *Bright Air Brilliant Fire: On the Matter of the Mind* 112, 115, 122–123 (Basic Books 1992).

60. Donald R. Griffin, *supra* note 1, at 211, quoting Carl Jung, *Synchronicity: A Causal Connecting Principle* 94 (Princeton University Press 1973).

61. Adrian M. Wenner and Patrick H. Wells, "Stinging criticism," 88 *American Scientist* 3, 3 (January-February 2000).

62. Mary Jo Nye, "N-rays: An episode in the history and psychology of science," 11 *Historical Strides in the Physical Sciences* 125 (1980).

63. Adrian M. Wenner and Patrick H. Wells, *Anatomy of a Controversy: The Question of a "Language" Among Bees* (Columbia University Press 1990); William Broad and Nicholas Wade, *Betrayers of the Truth: Fraud and Deceit in the Halls of Science* (Touchstone Books 1982).

64. Adrian M. Wenner and Patrick H. Wells, *supra* note 63, at 10.

65. *Id.* at 261.

66. Reply of Dr. Nieh, 88 *American Scientist* 3–4 (January-February 2000).

67. Michael Tomasello and Josep Call, *Primate Cognition* 174, 26 (Oxford University Press 1997). *See id.* at 173.

68. S. Glickman and R. Sroges, "Curiosity in zoo animals," 26 *Behavior* 161 (1966).

69. Michael Tomasello and Josep Call, *supra* note 67, at 172.

70. Barry O. Hughes and Arthur J. Black, "The preference of domestic hens for different types of battery cage floor," 14 *British Poultry Science* 615 (1973).

71. A. P. Silverman, "Rodents' defense against cigarette smoke," 26 *Animal Behavior* 1279 (1978).

72. Marc D. Hauser, *supra* note 20, at 223.

73. Marian Stamp Dawkins, "Battery hens name their price: Consumer demand theory and the measurement of ethological needs," 31 *Animal Behavior* 1195 (1983).

74. R. E. Snodgrass, *supra* note 7, at 248.

75. Personal communication from M. E. Bitterman, dated August 21, 2001.

76. Donald R. Griffin, *supra* note 1, at 268.

77. A. A. Milne, *Winnie the Pooh* (E.P. Dutton 1926). Reprinted in *The Book of Children's Classics* 225, 239 (Penguin Putnam 1997).

78. Website of the American Beekeeping Federation (http://www.abfnet.org/Industry_News/NASS2000CropReport.html) (visited August 15, 2001).

79. Website of British Beekeeping Association (http://www.bbka.org.uk/faq_honeybees.htm) (visited August 15, 2001).

80. Personal communication from Rick Bogle, dated August 15, 2001.

81. *E.g., State v. Cleve,* 980 P. 2d 23 (N.M. 1999) (wild game animals); *State v. Buford,* 65 N.M. 51 (1958) (fighting gamecocks); *State v. Claiborne,* 505 P. 2d 732 (Kan. 1973).

Chapter 6 Alex

1. Jeffrey Moussaieff Masson and Susan McCarthy, *When Elephants Weep: The Emotional Lives of Animals* 183 (Delacorte Press 1995).

2. *See* the discussion of language use by chimpanzees and bonobos in Steven M. Wise, *Rattling the Cage: Toward Legal Rights for Animals* 217–230 (Perseus Publishing 2000).

3. Juan Carlos Gómez, "The emergence of intentional communication as a problem-solving strategy in the gorilla," in *'Intelligence' and Language in Monkeys and Apes* 333, 345 (Sue Taylor Parker and Kathleen Rita Gibson, eds., Cambridge University Press 1990).

4. Juan Carlos Gómez, *supra* note 3, at 336–337, 346.

5. *E.g.*, Irene Maxine Pepperberg, *The Alex Studies: Cognitive and Communicative Abilities of Grey Parrots* 5, 34, 95 (Harvard University Press 1999); Irene M. Pepperberg *et al.*, "Solitary sound play during acquisition of English vocalizations by an African Grey parrot *(Psittacus erithacus):* Possible parallels with children's monologue speech," 12 *Applied Psycholinguistics* 151, 154–155 (1991); Irene Maxine Pepperberg, "Conceptual abilities of some nonprimate species, with an emphasis on an African Grey parrot," in Sue Taylor Parker and Kathleen Rita Gibson, eds., *supra* note 3, at 469, 474, 498.

6. John C. Marshall, "Not just a pretty Polly," 404 *Nature* 929, 930 (April 27, 2000).

7. Irene Maxine Pepperberg *(The Alex Studies), supra* note 5, at 130.

8. *Id.* at 198, 12.

9. *Id.* at 10, 19, 156. 195, 322.

10. Donald R. Griffin, *Animal Minds: Beyond Cognition to Consciousness* 186 (University of Chicago Press 2001).

11. Irene Maxine Pepperberg *(The Alex Studies), supra* note 5, at 208; Irene M. Pepperberg, "Rethinking syntax: A commentary on E. Kako's 'Elements of syntax in the systems of three language-trained animals,'" 27 *Animal Learning & Behavior* 15, 17 (1999);

Richard Byrne, *The Thinking Ape: Evolutionary Origins of Intelligence* 174 (Oxford University Press 1995).

12. Irene Maxine Pepperberg *(The Alex Studies), supra* note 5, at 212, 217, 222, 225, 226; Irene M. Pepperberg *et al., supra* note 5, at 163–172.

13. Donald R. Griffin, *supra* note 10, at 183.

14. David L. Chandler, "This bird talks, counts, and reads—a little," *Boston Globe,* at D1, D3 (May 18, 1998).

15. David Abel, "Web-savvy parrots fight pet boredom," *Boston Globe,* at D1, D3 (September 26, 2000).

16. Joanna Burger, *The Parrot Who Owns Me: The Story of a Relationship* 5 (Villard 2001).

17. Personal communication from "Mr. Darwin," dated June 10, 2001.

18. Personal communication from Irene M. Pepperberg, dated February 10, 2001.

19. Irene Maxine Pepperberg *(The Alex Studies), supra* note 5, at 8–10; Sue Taylor Parker and Kathleen Rita Gibson, eds., *supra* note 3, at 474–475.

20. Irene Maxine Pepperberg *(The Alex Studies), supra* note 5, at 9.

21. *E.g., id.* at 236; Irene M. Pepperberg, "Referential mapping: A technique for attaching functional significance to the innovative utterances of an African Grey parrot *(Psittacus erithacus),*" 11 *Applied Psycholinguistics* 22, 29 (1990).

22. Irene Maxine Pepperberg *(The Alex Studies), supra* note 5, at 2–20.

23. Irene M. Pepperberg, "The importance of social interaction and observation in the acquisition of communicative competence: Possible parallels between avian and human learning," in *Social Learning: Psychological and Biological Perspectives* 279, 279–280 (Thomas R. Zentall and Bennett G. Galef, Jr., eds., Lawrence Erlbaum, 1988).

24. Irene Maxine Pepperberg *(The Alex Studies), supra* note 5, at 24–29; Irene M. Pepperberg, "Social interaction as a condition for learning in avian species: A synthesis of the disciplines of ethology and psychology," in *The Inevitable Bond* 178, 192–193 (Hank Davis and David Balfour, eds., Cambridge University Press 1992).

25. Irene M. Pepperberg, *supra* note 23, at 281.

26. Irene Maxine Pepperberg *(The Alex Studies), supra* note 5, at 20–22.

27. Irene Maxine Pepperberg *(The Alex Studies), supra* note 5, at 264–268; Irene M. Pepperberg, "Effect of avian-human joint attention on allospecific vocal learning by grey parrots *(Psittacus erithacus),*" 110 *Journal of Comparative Psychology* 286, 292–295 (1996); Irene M. Pepperberg, "Vocal learning in grey parrots *(Psittacus erithacus):* Effects of social interaction, reference, and context," 111 *The Auk* 300, 309–311 (1994); Irene M. Pepperberg, "Social modeling theory: A possible framework for understanding avian vocal learning," 102 *The Auk* 854, 854–858 (1985).

28. Irene Maxine Pepperberg *(The Alex Studies), supra* note 5, at 32.

29. *E.g.,* Irene M. Pepperberg *et al.,* "Vocal learning in the grey parrot *(Psittacus erithacus):* Effects of species identity and number of trainers," 114 *Journal of Comparative Psychology* 371 (2000); Irene M. Pepperberg *et al.,* "Limited contextual vocal learning in the grey parrot *(Psittacus erithacus):* The effect of interactive co-viewers on videotaped instruction," 113 *Journal of Comparative Psychology* 158 (1999).

30. Irene Maxine Pepperberg *(The Alex Studies), supra* note 5, at 10.

31. Irene Maxine Pepperberg *(The Alex Studies), supra* note 5, at 24–29; Irene M. Pepperberg, *supra* note 24, at 188–192; Irene Pepperberg, "A communicative approach to an-

imal cognition: A study of conceptual abilities of an African Grey parrot," in *Cognitive Ethology: The Minds of Other Animals* 153, 161 (Carolyn Ristau, ed., Lawrence Erlbaum 1991).

32. Bruce R. Moore, "Avian movement imitation and a new form of mimicry: Tracing the evolution of a complex form of behavior," 122 *Behavior* 231, 241–256 (1992).

33. Robert W. Mitchell, "Anthropomorphism and anecdotes: A guide for the perplexed," in *Anthropomorphism, Anecdotes, and Animals* 418–419 (Robert W. Mitchell *et al.*, eds., State University of New York Press 1997); H. Lyn Miles, "Anthropomorphism, Apes, and Language," in Robert W. Mitchell *et al.,* eds., *supra* note 23, at 383, 392.

34. Jerome Bruner, "Tot thought," XLVII(4) *New York Review of Books* 27, 28 (March 9, 2000).

35. Claudia Dreifus, "Do androids dream? M.I.T. is working on it," *New York Times* (November 7, 2000) (emphases added).

36. I thank Anne Russon, an expert on imitation, for carefully reviewing and improving my discussion of imitation, Personal communication from Anne Russon, dated February 17, 2001.

37. Richard W. Byrne and Anne E. Russon, "Learning by imitation: A hierarchical approach," in 21 *Behavioral and Brain Sciences* 667, 667 (1998); Richard W. Byrne, "The misunderstood ape: Cognitive skills of the gorilla," in *Reaching Into Thought: The Minds of the Great Apes* 111, 123 (Anne E. Russon *et al.,* eds., Cambridge University Press 1996); Richard Byrne, *supra* note 11, at 65; H. Lyn Miles *et al.,* "Simon says: The development of imitation in an enculturated orangutan," in Anne E. Russon *et al.,* eds., *supra* at 278, 280.

38. Alison Gopnik and Andrew Meltzoff, "Minds, bodies, and persons: Young children's understanding of the self and others as reflected in imitation and theory of mind research," in S*elf-Awareness in Animals and Humans: Developmental Perspectives* 166 (Sue Taylor Parker *et al.,* eds., Cambridge University Press 1994).

39. *E.g.*, H. Lyn Miles *et al., supra* note 37, at 280–181, 293; Daniel Hart and Mary Pat Karmel, "Self-awareness and self-knowledge in humans, apes, and monkeys," in Anne E. Russon *et al.,* eds., *supra* note 37, at 325, 335; Daniel Hart and Suzanne Fegley, "Social imitation and a mental model of self," in Sue Taylor Parker *et al., supra* note 22, at 149–162. *But see* Deborah Custance and Kim A. Bard, "The comparative and developmental study of self-recognition and imitation: The importance of social factors," in Sue Taylor Parker *et al.,* eds., *supra* note 22, at 207.

40. Andrew Whiten, "Chimpanzee cognition and the question of mental re-representation," in *Metarepresentation: A Multidisciplinary Perspective* 139, 144–145 (Daniel Sperber, ed., Oxford University Press 2000); Alison Gopnik *et al.,* "Early theories of mind: What the theory theory can tell us about autism," in *Understanding Other Minds: Perspectives from Developmental Cognitive Neuroscience* 50, 59–61 (Simon Baron-Cohen *et al.,* eds., 2d ed., Oxford University Press 2000); Andrew Whiten, "Imitation, pretense, and mindreading: Secondary representation in comparative primatology and developmental psychology?" in Anne E. Russon *et al.,* eds., *supra* note 37, at 300, 308–309.

41. Alison Motluk, "Read my mind," 169 *New Scientist* 21, 21–26 (January 27, 2001). *See* Vittorio Gallese and Alvin Goldman, "Mirror neurons and the simulation theory of mind-reading," 2 *Trends in Cognitive Neuroscience* 493 (1998); Giacomo Rizzolatti and Michael Arbib, "Language within our grasp," 21 *Trends in Neurosciences* 188 (1998).

42. A. L. Kroeber and C. Kluckhohn, "Culture, a critical review of the concepts and definitions," 47 *American Archaeology and Ethnology* 1 (1952).

43. Daniel C. Dennett, *Darwin's Dangerous Idea: Evolution and the Meanings of Life* 361–368 (Simon and Schuster 1995); Richard Dawkins, *The Selfish Gene* 206 (Oxford University Press 1976). Others disagree, *e.g.*, Robert Boyd and Peter J. Richerson, "Meme theory oversimplifies how culture changes," 283 *Scientific American* 70, 70–71 (October 2000).

44. Susan Blackmore, *The Meme Machine* 43 (Oxford University Press 1999). Others argue that memes can spread in other ways, Henry Plotkin, "People do more than imitate," 283 *Scientific American* 72, 72 (October 2000). For in-depth debates about the science of memetics, *see Darwinizing Culture* (Robert Aunger, ed., Oxford University Press 2001).

45. Susan Blackmore, *supra* note 44, at 50.

46. *Id.* at 52.

47. Lee Alan Dugatkin, *The Imitation Factor* 131 (Free Press 2000).

48. *Id.* at 119; Susan Blackmore, "The power of memes," 283 *Scientific American* 64, 65 (October 2000).

49. Susan Blackmore, *supra* note 44, at 7. *See* Susan Blackmore, *supra* note 48, at 65.

50. Lee Alan Dugatkin, *supra* note 47, generally; Lee Alan Dugatkin, "Animals imitate, too," 283 *Scientific American* 64, 65, 67 (October 2000). *See* Andrew Whiten and Christophe Boesch, "The cultures of chimpanzees," 284(1) *Scientific American* 61 (January 2001); A. Whiten *et al.*, "Cultures in chimpanzees," 399 *Science* 682, 682, 686 (1999); Blackmore has acknowledged that many birds, whales, dolphins, and chimpanzees imitate, Susan Blackmore, *supra* note 48, at 68.

51. Edward L. Thorndike, *Animal Intelligence* 76, 78–79 (Macmillan 1911). *See* Richard W. Byrne and Anne E. Russon, *supra* note 37, at 668; Andrew Whiten and Rebecca Ham, "On the nature and evolution of imitation in the animal kingdom: Reappraisal of a century of research," 21 *Advances in the Study of Behavior* 239, 240 (1992).

52. J. Fisher and Robert Hinde, "The opening of milk bottles by birds," 44 *British Birds* 393 (1951).

53. David. F. Sherry and Bennett G. Galef, "Cultural transmission without imitation: Milk bottle opening by birds," 32 *Animal Behaviour* 937, 937–938 (1984). *See* Richard W. Byrne and Anne E. Russon, *supra* note 37, at 669; Richard Byrne, *supra* note 11, at 56–58.

54. Richard W. Byrne and Anne E. Russon, *supra* note 37, at 668, 677.

55. Michael Tomasello *et al.*, "Observational learning of tool-use by young chimpanzees," 2 *Human Evolution* 175 (1987). *Compare* Josep Call, "Levels of imitation and cognitive mechanisms in orangutans," in Sue Taylor Parker *et al.*, eds., *supra* note 13, at 316, 317 ("emulation") with Andrew Whiten and Rebecca Ham, *supra* note 51, at 251–252 ("goal emulation"). *See* Richard W. Byrne and Anne E. Russon, *supra* note 37, at 669; Richard Byrne, *supra* note 11, at 59–60.

56. Personal communication from Anne Russon, dated February 17, 2001. Richard W. Byrne and Anne E. Russon, *supra* note 37, at 676.

57. Andrew Whiten, "Primate culture and social learning," 24 *Cognitive Science* 477, 479–483 (2000).

58. Richard W. Byrne, *supra* note 37, at 124.

59. Joanna Burger, *supra* note 16, at 44.

60. Ina C. Uzgiris and J. McV. Hunt, *Assessment in Infancy: Ordinal Scales of Psychological Development* 151–164 (University of Illinois Press 1975).

61. Irene Maxine Pepperberg *(The Alex Studies), supra* note 5, at 169.

62. *Id.* at 179; Irene M. Pepperberg, "Development of Piagetian object permanence in a grey parrot *(Psittacus erithacus),*" 111 *Journal of Comparative Psychology* 63, 65–66 (1997).

63. Personal communication from Irene Pepperberg, dated August 1, 2001; Irene M. Pepperberg, *supra* note 62, at 69–70; Irene M. Pepperberg and Florence A. Kozak, "Object permanence in the African Grey parrot," 14(3) *Animal Learning and Behavior* 322, 326 (1986).

64. Irene Pepperberg, "Ruffled feathers," in *The Smile of a Dolphin: Remarkable Accounts of Animal Emotions* 108 (Marc Bekoff, ed., Discovery Books 2000); Irene Maxine Pepperberg *(The Alex Studies), supra* note 5, at 171; Irene M. Pepperberg, *supra* note 62, at 70 and note 8.

65. Irene Maxine Pepperberg *(The Alex Studies), supra* note 5, at 186. Some food-storing birds pass most of the standard Stage 6 tests, Bettina Politick, "Development of object permanence in food-storing magpies *(Pica pica),*" 114 *Journal of Comparative Psychology* 148 (2000). Other birds attain Stage 4 competency in object permanence tasks, but no higher, Claude Dumas and Donald M. Wilkie, "Object permanence in ring doves *(streptopelia risoria),*" 109 *Journal of Comparative Psychology* 142 (1995), while some fail Stage 4 tasks, C. M. S. Plowright *et al.*, "Finding hidden food: Behavior on visible displacement tasks for mynahs *(Gracula religiosa)* and pigeons *(Columba livia),*" 112 *Journal of Comparative Psychology* 13 (1998) (pigeons), or pass Stage 4 tests without being tested on higher levels, *id.* (mynahs).

66. Irene Maxine Pepperberg *(The Alex Studies), supra* note 5, at 46.

67. Thomas Bugnyar *et al.*, "Food calling in ravens: Are yells referential signals?" 61 *Animal Behaviour* 949, 956 (2001); Dorothy L. Cheney and Robert M. Seyfarth, *How Monkeys See the World* 102–110 (University of Chicago Press 1990).

68. Irene Maxine Pepperberg *(The Alex Studies), supra* note 5, at 45–47; Irene M. Pepperberg, "Cognition in the African Grey parrot: Preliminary evidence for auditory/vocal comprehension of the class concept," 11(2) *Animal Learning & Behavior* 179, 179–180 (1983).

69. Irene Maxine Pepperberg *(The Alex Studies), supra* note 5, at 43, 235–236; Irene M. Pepperberg, *supra* note 24, at 190–192; Irene M. Pepperberg, *supra* note 21, at 25–31.

70. Irene Maxine Pepperberg *(The Alex Studies), supra* note 5, at 52–54; Irene M. Pepperberg, *supra* note 68, at 179–180.

71. Irene Pepperberg, *supra* note 24, at 165.

72. Irene Maxine Pepperberg *(The Alex Studies), supra* note 5, at 57; Irene M. Pepperberg, *supra* note 68, at 181.

73. Irene Maxine Pepperberg *(The Alex Studies), supra* note 5, at 61; Irene M. Pepperberg, *supra* note 68, at 183.

74. Irene Maxine Pepperberg *(The Alex Studies), supra* note 5, at 62–64; Sue Taylor Parker and Kathleen Rita Gibson, eds., *supra* note 3, at 490; Irene M. Pepperberg, "Acquisition of the same-different concept by an African Grey parrot *(Psittacus erithacus):* Learning with respect to categories of color, shape, and material," 15(4) *Learning & Behavior* 423, 423–424 (1987).

75. Irene Maxine Pepperberg *(The Alex Studies), supra* note 5, at 66.

76. *Id.* at 73; Sue Taylor Parker and Kathleen Rita Gibson, eds., *supra* note 3, at 495–497; Irene Pepperberg, *supra* note 31, at 167–178; Irene M. Pepperberg, *supra* note 74, at 428.

77. Irene Maxine Pepperberg *(The Alex Studies), supra* note 5, at 81.

78. *Id.*

79. *Id.* at 86–94; Irene M. Pepperberg, "Comprehension of 'absence' by an African Grey parrot: Learning with respect to questions of same/different," 50 *Journal of the Experimental Analysis of Behavior* 553, 560–563 (1988).

80. Irene Maxine Pepperberg *(The Alex Studies), supra* note 5, at 160–166; Irene M. Pepperberg and Michael V. Brezinsky, "Acquisition of a relative class concept by an African Grey parrot *(Psittacus erithacus):* Discriminations based on relative size," 105 *Journal of Comparative Psychology* 286 (1991). Pepperberg used other objects.

81. Irene Maxine Pepperberg *(The Alex Studies), supra* note 5, at 164–165.

82. Irene Maxine Pepperberg *(The Alex Studies), supra* note 5, at 127; Irene M. Pepperberg, *supra* note 21, at 41.

83. Irene Maxine Pepperberg *(The Alex Studies), supra* note 5, at 128–129.

84. Irene M. Pepperberg, "Proficient performance of a conjunctive, recursive task by an African Grey parrot *(Psittacus erithacus),*" 106 *Journal of Comparative Psychology* 295, 301–302 (1992).

85. Irene Maxine Pepperberg *(The Alex Studies), supra* note 5, at 198.

86. *Id.*

87. *Id.* at 200; Irene M. Pepperberg, "An interactive modeling technique for acquisition of communication skills: Separation of 'labeling' and 'requesting' in a psittacine subject," 9 *Applied Psycholinguistics* 59, 65 (1988).

88. Irene Maxine Pepperberg *(The Alex Studies), supra* note 5, at 201–202; Irene M. Pepperberg, *supra* note 87, at 66.

89. Irene Maxine Pepperberg *(The Alex Studies), supra* note 5, at 205.

90. Irene Maxine Pepperberg *(The Alex Studies), supra* note 5, at 201, 202, 203; Irene M. Pepperberg, *supra* note 87, at 67, 69.

91. Jeffrey Moussaieff Masson and Susan McCarthy, *supra* note 1, at 19.

92. Personal communication from Irene Pepperberg, dated February 7, 2001.

93. Jeffrey Moussaieff Masson and Susan McCarthy, *supra* note 1, at 229.

94. Irene Maxine Pepperberg *(The Alex Studies), supra* note 5, at 245.

95. *Id.* at 203, 246

96. *Id.* at 340, note 1; Irene M. Pepperberg, *supra* note 87, at 73, note 3.

97. Irene Maxine Pepperberg *(The Alex Studies), supra* note 5, at 206.

98. *Id.* at 203; Irene M. Pepperberg, *supra* note 87, at 70; Irene M. Pepperberg, "Interspecies communication: A tool for assessing conceptual abilities in the African Grey parrot *(Psittacus erithacus),* in *Cognition, Language, and Consciousness* 31 (Gary Greenberg and Ethel Tobach, eds., Lawrence Erlbaum 1987).

99. Irene Maxine Pepperberg *(The Alex Studies), supra* note 5, at 243–244.

100. *Id.* at 244; Irene M. Pepperberg, *supra* note 21, at 33.

101. Irene Maxine Pepperberg *(The Alex Studies), supra* note 5, at 61, 244, 334, note 4; Irene M. Pepperberg, *supra* note 21, at 39.

102. Jeffrey Moussaieff Masson and Susan McCarthy, *supra* note 1, at 35.

103. Eugene Linden, *The Parrot's Lament: And Other True Tales of Animal Intrigue, Intelligence, and Ingenuity* 40 (Dutton Books 1999).

104. Donald R. Griffin, *supra* note 7, at 174.

105. Richard Byrne, *supra* note 11, at 173.

106. Irene Maxine Pepperberg *(The Alex Studies), supra* note 5, at 100.

107. *Id.* at 116; Irene M. Pepperberg, "Numerical competence in an African Grey parrot *(Psittacus erithacus),*" 108 *Journal of Comparative Psychology* 36, 39 (1994).

108. Irene Maxine Pepperberg *(The Alex Studies), supra* note 5, at 120–121.

109. *Id.* at 123–124. *See* Irene M. Pepperberg, *supra* note 107, at 41–42; Sue Taylor Parker and Kathleen Rita Gibson, eds., *supra* note 3, at 479–489.

110. Irene Maxine Pepperberg *(The Alex Studies), supra* note 5, at 34.

111. *Id.* at 58, 60, 61, 75–76, 79, 84, 88, 94, 125, 137–141, 149–151, 166–168, 192.

112. Bettinia Pollok *et al.,* "Development of object permanence in food-storing magpies *(Pica pica),*" 114 *Journal of Comparative Psychology* 148 (2000).

113. Irene M. Pepperberg, "Object permanence in four species of psittacine birds: An African Grey parrot *(Psittacus erithacus),* an Illiger mini macaw *(Ara maracana),* a parakeet *(Melopsittacus undulatus),* and a cocakatiel *(Nymphicus hollandicus),* 18 *Animal Learning and Behavior* 97, 100–102 (1990) (The parakeet passed Task 13 before he died.)

114. *E.g.,* Nicola S. Clayton and Anthony Dickenson, "Episodic-like memory during cache recovery by scrub jays," 395 *Nature* 272 (1998). *See* Kristy Gould-Beierle, "A comparison of four corvid species in a working and reference memory task using a radial maze," 14 *Journal of Comparative Psychology* 347 (2000).

115. Joanna Burger, *supra* note 16, at 137.

116. Nicola S. Clayton *et al.,* "Declarative and episodic-like memory in animals: Personal musings of a scrub jay," in *The Evolution of Cognition* 273, 273 (Cecelia Heyes and Ludwig Huber, eds., MIT Press 2000); Daniel Griffiths *et al.,* "Episodic memory: What can animals remember about their past?" 3 *Trends in Cognitive Science* 74, 76–80 (1999); Nicola S. Clayton and Anthony Dickinson, *supra* note 114, at 274.

117. Nicola S. Clayton *et al.,* "Scrub jays *(Aphelocoma coerulescens)* for integrated memories of the multiple features of caching episodes," 27(1) *Journal of Experimental Psychology: Animal Behavior Processes* 17, 27–28 (2001); Nicola S. Clayton *et al., supra* note 114, at 278–286; Nicola S. Clayton and Anthony Dickinson, "Scrub jays *(Aphelocoma coerulescens)* remember the relative time of caching as well as the location and content of their caches," 113 *Journal of Comparative Psychology* 403, 413 (1999); Nicola S. Clayton and Anthony Dickinson, "Memory for the content of caches by scrub jays *(Aphelocoma coerulescens),*" 25(1) *Journal of Experimental Psychology: Animal Behavior Processes* 82, 88–90 (1999); Nicola S. Clayton and Anthony Dickinson, *supra* note 114, at 274; Kathryn Jeffrey and John O'Keefe, "Worm holes and avian space-time," 395 *Nature* 215, 215 (September 17, 1998).

118. Gavin R. Hunt, "Manufacture and use of hook-tools by New Caledonian crows," 379 *Nature* 249 (January 18, 1996). *See* Christophe Boesche, "The question of culture," 379 *Nature* 207 (January 18, 1995).

119. Gavin R. Hunt, *et al.* "Laterality in Tool Manufacture by Crows," 414 *Nature* 707, 707 (2001).

120. Thomas Bugnyar *et al., supra* note 67, at 956–957; Bernd Heinrich, *Mind of the Raven* xix (HarperCollins 1999).

121. Bernd Heinrich, "Hopping mad," in Marc Bekoff, ed., *supra* note 64, at 98–99; Bernd Heinrich, *supra* note 120, at 45, 191, 341, 348.

122. *Id.* at 31–48.

123. *Id.* at 31.

124. Sy Montgomery, "The raven: Back from the brink of nevermore," *Boston Globe* at C1, C5 (March 30, 1999).

125. Personal communication from Bernd Heinrich, dated March 15, 2001; Bernd Heinrich and John W. Pepper, "Influence of competitors on caching behaviour in the common raven, *Corvus corax*," 56 *Animal Behaviour* 1083, 1089 (1998).

126. Bernd Heinrich, "Planning to facilitate caching: Possible suet cutting by a common raven," 111 *Wilson Bulletin* 296, 297–299 (1999); Bernd Heinrich, *supra* note 120, at 268, 310–311.

127. Bernd Heinrich, "Testing insight in ravens," in *The Evolution of Cognition* 289, 292–299 (Cecelia Heyes and Ludwig Huber, eds., MIT Press 2000).

128. Bernd Heinrich, *supra* note 126, at 296–297; Bernd Heinrich, *supra* note 120, at 314, 317–321. *See* Bernd Heinrich, "An experimental investigation of insight in common ravens *(Corvus corvax)*," 112 *The Auk* 994, 995–1002 (1885).

129. Bernd Heinrich, *supra* note 126, at 299–303; Bernd Heinrich, *supra* note 120, at 354–356.

130. Bernd Heinrich, *supra* note 120, at 331.

131. Douglas Chadwick, "Ravens: Legendary bird brains," 195 *National Geographic* 100, 108 (January 1999).

132. Joanna Burger, *supra* note 16, at 44.

133. *Id.* at 5.

134. Bernd Heinrich, *supra* note 120, at 176.

Chapter 7 Marbury

1. Carles Vila *et al.*, "Multiple and ancient origins of the domestic dog," 276 *Science* 1687 (1997).

2. Krisztina Soproni *et al.*, "Comprehension of human communicative signs in pet dogs *(Canis familiaris)*," 115(2) *Journal of Comparative Psychology* 122, 122 (2001).

3. Peter Pongrácz *et al.*, "Owners' beliefs on the ability of their pet dogs to understand verbal communication: A case of social understanding," 20 *Current Psychology of Cognition* 87, 90 (2001).

4. Carles Vila *et al.*, *supra* note 1.

5. Personal communication from "Mr. Darwin," dated May 20, 2001.

6. Elizabeth McKey and Karen Payne, "APPMA Study: Pet ownership soars," 18(8) *Pet Business* 22 (August 1992).

7. Jozsef Topal *et al.*, "Attachment behavior in dogs *(Canis familiaris):* New application of Ainsworth's (1969) Strange Situation test," 112(3) *Journal of Comparative Psychology* 219, 220 (1998).

8. *Id.* at 225–226 (1998).

9. Charles Darwin, *On the Origin of Species* 215 (facsimile of first edition, Harvard University Press 1964).

10. Krisztina Soproni *et al.*, *supra* note 2, at 122.

11. H. Frank, *et al.*, "Motivation and insight in wolf *(Canis lupus)* and Alaskan malamute *(Canis familiaris):* Visual discrimination learning," 27(5) *Bulletin of Psychonomic Society* 455, (1989); H. Frank and M. G. Frank, "Comparative manipulation test performance in ten-week-old wolves *(Canis lupus)* and Alsatian malamutes *(Canis familiaris):* A Piagetian interpretation," 99(3) *Journal of Comparative Psychology* 266 (1985); H. Frank and M. G. Frank, "Comparison of problem-solving performance in six-week-old wolves and dogs," 30(1) *Animal Behaviour* 95, 97–98 (1982).

12. Jozsef Topal *et al.*, "Dog-human relationship affects problem-solving behavior in the dog," 10(4) *Anthrozoos* 214, 215–222 (1997);

13. Brian Hare (unpublished manuscript).

14. Personal communication from Brian Hare, dated July 2001.

15. Daniel C. Dennett, *Kinds of Minds: Toward an Understanding of Consciousness* 165 (Basic Books 1996). In *Rattling the Cage,* I criticized Dennett for what I understood to be his claim that no nonhumans have minds. As is my custom, I sent Dennett drafts of my criticism of his work and asked for his comments. He was too busy to accept my offer. After publication, he contacted me to say that I had distorted his views. He says he believes that some nonhuman animals have minds, though they differ from human minds, and that not only do nonhuman animals suffer less than humans but they suffer differently. He also believes that the consciousness of human children is much less similar to human adult consciousness than many believe. He says we agree that behavioral evidence, analyzed in a sophisticated way, is the key to understanding nonhuman animal minds. And he denies believing that either computer programs or robots are conscious. To the extent I misunderstood your writings, Professor, *mea culpa,* and I am delighted that our views are so much closer than I believed. Personal communication from Daniel C. Dennett, dated March 4, 2001.

16. Homer, *The Odyssey,* Book 17.317–334, 359–360, at 363–364 (Robert Fagles, trans., Viking 1996).

17. Richard Sorabji, *Animal Minds and Human Morals: The Origins of the Western Debate* 21 (Cornell University Press 1993).

18. Elizabeth Marshall Thomas, *The Hidden Life of Dogs* 57–58 (Houghton Mifflin 1993). Wolves are much more clearly monogamous and have favorite mates over which they become jealous, S. Pal *et al.*, "Inter- and intra-sexual behaviour of free-ranging dogs *(Canis familiaris),* 62 *Applied Animal Behavioural Science* 267 (1999). I thank Brian Hare for bringing this study to my attention.

19. Roger Fouts and Stephen Tukel Mills, *Next of Kin: What Chimpanzees Have Taught Me About Who We Are* 6 (William Morrow and Co. 1997).

20. Jeffrey Moussaieff Masson, *Dogs Never Lie About Love* xxi (Crown Publishers 1997).

21. Marc Bekoff, review of Rupert Sheldrake's *Dogs That Know When Their Owners Are Coming Home and Other Unexplained Powers of Animals,*" 10 *Bark* 44 (2000); Rupert Sheldrake, *Dogs That Know When Their Owners Are Coming Home and Other Unexplained Powers of Animals* (Crown Publishers 1999).

22. Michael W. Fox, "The nature of compassion," in *The Smile of the Dolphin: Remarkable Accounts of Animal Emotions* 177, 178 (Marc Bekoff, ed., Discovery Books 2000).

23. Clinton R. Sanders, "Simple pleasures," in Marc Bekoff, ed., *supra* note 22, at 128, 128.

24. *Id.*

25. Stanley Coren, *The Intelligence of Dogs: Canine Consciousness and Capabilities* 12 (The Free Press 1994).

26. *Id.*

27. Dorit Urd Feddersen-Petersen, "Vocalization of European wolves *(Canis lupus lupus L.)* and various dog breeds *(Canis lupus f. fam),* 43 *Archives of Animal Breeding* 387, 388 (2000).

28. Marc Bekoff, "Demonic dogs: The alleged 'truth' about would-be jerks," *review of The Truth About Dogs: An Inquiry Into the Ancestry, Social Conventions, Mental Habits, and Moral Fiber of* Canis familiaris," 14(1) *Anthrozoos* 56, 56 (2001)

29. *Id.* at 57.

30. Personal communication from Brian Hare, dated July 2001.

31. James R. Anderson, "The monkey in the mirror: A strange conspecific," in Sue Taylor Parker *et al., supra* note 38, at 315, 318.

32. Jesheskel Shoshani, "It's a nose! It's a hand! It's an elephant's trunk!" 106 *Natural History* 36, 37 (November 1997); Mark Derr, *Dog's Best Friend: Annals of the Dog-Human Relationship* 95 (Henry Holt 1997).

33. Antonio Damasio, *The Feeling of What Happens* 198 (Harcourt Brace 1999) (emphasis added).

34. Emanuela Cenami Spada *et al., supra* note 87, at 196, 210.

35. Marc D. Hauser, *Wild Minds: What Animals Really Think* 76 (Henry Holt 2000); Nicole Chapuis and Christian Varlet, "Short cuts by dogs in natural surroundings," 39B *The Quarterly Journal of Experimental Psychology* 49, 51–63 (1987).

36. Donald R. Griffin, *Animal Thinking* 203 (Harvard University Press 1984).

37. Kenway Louis and Matthew A. Wilson, "Temporally structured replay of awake hippocampal ensemble activity during rapid eye movement sleep," 29 *Neuron* 145, 145–154 (2001).

38. Estella Triana and Robert Pasnak, "Object permanence in cats and dogs," 9 *Animal Learning and Behavior* 135, 137–138 (1981).

39. Sylvain Gagnon and François Y. Doré, "Search behavior in various breeds of adult dogs *(Canis familiaris):* Object permanence and olfactory cues," 106(1) *Journal of Comparative Psychology* 58, 58–59 (1992). Unlike humans and apes, puppies did not make the "A-not-B" errors at Stage 4, Sylvain Gagnon and François Y. Doré, "Cross-sectional study of object permanence in domestic puppies *(Canis familiaris),*" 108 *Journal of Comparative Psychology* 220, 227 (1994).

40. Michael Tomasello and Josep Call, *Primate Cognition* 41 (Oxford University Press 1997); Sylvain Gagnon and François Y. Doré (1994), *supra* note 39.

41. Sylvain Gagnon and François Y. Doré (1992), *supra* note 39, at 66; Sylvain Gagnon and François Y. Doré (1994), *supra* note 39, at 230.

42. Sylvain Gagnon and François Y. Doré, "Search behavior of dogs *(Canis familiaris)* in invisible displacement problems," 21(3) *Animal Learning & Behavior* 246, 246–247 (1993); Sylvain Gagnon and François Y. Doré (1992) *supra* note 39, at 60–63, 67.

43. François Y. Doré *et al.,* "Search behavior in cats and dogs: Interspecific differences in working memory and spatial cognition," 24(2) *Animal Learning & Behavior* 142, 142–143 (1996); Sylvain Gagnon and François Y. Doré (1992) *supra* note 39, at 67.

44. François Y. Doré *et al., supra* note 43, at 143.

45. *Id.* at 148.

46. François Y. Doré and Sonia Goulet (1992) *supra* note 39, at 59–60.

47. Jeffrey Moussaieff Masson, *supra* note 20, at 29–30 (Masson's statement that "I know of no contemporary research having been done on memory in dogs," predates Gagnon and Dore's work).

48. *E.g.*, *id.* at 30–33.

49. Lesley J. Rogers, *Minds of Their Own: Thinking and Awareness in Animals* 55 (Westview Press 1997).

50. Claudia Dreifus, "Observing the behavior of apes, up close," *New York Times*, at D3 (June 26, 2001).

51. Joanna Burger, *The Parrot Who Owns Me: The Story of a Relationship* 43–44 (Villard Books 2001).

52. Ruth G. Millikan, *Language, Thought, and Other Biological Categories* 85–113 (MIT Press 1984).

53. Marc Bekoff, "Playing with play: What can we learn about cognition, negotiation, and evolution?" in *The Evolution of Mind* 162, 163, 166 (Denise Dellarosa Cummins and Colin Allen, eds., Oxford University Press 1998); Preliminary results from a new study suggest that dogs may posess at least a rudimentary theory of mind, Alexandra Horowitz, "Playing into awareness: Does a dog have a theory of mind?" (unpublished manuscript, University of California, San Diego).

54. Marek Spinka *et al.*, "Mammalian play: Training for the unexpected," 76(2) *The Quarterly Review of Biology* 141, 145 (June 2001).

55. Colin Allen and Marc Bekoff, *Species of Mind: The Philosophy and Biology of Cognitive Ethology* 89–98 (MIT Press 1997).

56. Krisztina Soproni *et al.*, *supra* note 2, at 122; Katie Douglas, "Mind of a dog," *New Scientist* 23, 24 (March 4, 2000).

57. Bryan Agnetta *et al.*, "Cues to food location that domestic dogs *(Canis familiaris)* of different ages do and do not use," 3 *Animal Cognition* 107, 110 (2000).

58. Mark Derr, "What do those barks mean? To dogs, it's all just talk?" *New York Times* at D5 (April 24, 2001); Dorit Urd Feddersen-Petersen, *supra* note 27, at 390.

59. Dorit Urd Feddersen-Petersen, *supra* note 27, at 391–395.

60. Mark Derr, *supra* note 58, at D5.

61. *Id.*

62. Brian Hare *et al.*, "Communication of food location between human and dog," 2(1) *Evolution of Communication* 137, 138 (1998); C. J. Warden and L. H. Warner, "The sensory capacities and intelligence of dogs, with a report on the ability of the noted dog "Fellow" to respond to verbal stimuli," 3(1) *The Quarterly Review of Biology* 1, 18–19 (1928).

63. Stanley Coren, *supra* note 25, at 93, 97.

64. *Id.* at 97, 114.

65. Peter Pongracz *et al.*, *supra* note 3, at 3.

66. *Id.* at 5.

67. *Id.* at 12–17.

68. *Id.* at 14.

69. Katie Douglas, *supra* note 56, at 26.

70. Stanley Coren, *supra* note 25, at 114.

71. C. J. Warden and L. H. Warner, *supra* note 62, at 1, 19.

72. *Id.* at 17.

73. *Id.* at 19.

74. *Id.* at 26.

75. Patricia B. McConnell and Jeffrey R. Baylis, "Interspecific communication in cooperative herding: Acoustic and visual signals from human shepherd and herding dogs," 67 *Zeitschrift für Tierpsychologie* 302, 314, 318, 321 (1985).

76. A. Miklósi *et al.,* "Use of experimenter-given cues in dogs," 1 *Animal Cognition* 113, 114 (1998).

77. *Id.* at 116–120.

78. Brian Hare *et al., supra* note 62, at 141–150, 155.

79. *Id.* at 154.

80. *Id.* at 155, 156.

81. Brian Hare and Michael Tomasello, "Domestic dogs *(Canis familiaris)* use human and conspecific social cues to locate food," 113(2) *Journal of Comparative Psychology* 173, 175–176 (1999).

82. *Id.*

83. *Id.*

84. *Id.*

85. Bryan Agnetta *et al., supra* note 57, at 108.

86. *Id.* at 112.

87. *Id.* at 108–110, 112.

88. Personal communication from Brian Hare, dated July 2001 (emphasis in the original).

89. A. Miklósi *et al.,* "Intentional behavior in dog-human communication: An experimental analysis of 'showing' behaviour in the dog," 3 *Animal Cognition* 159, 159 (2000).

90. *Id.* at 162–164.

91. Krisztina Soproni *et al., supra* note 2, at 122; A. Miklósi *et al., supra* note 89, at 165.

92. A. Miklósi *et al., supra* note 89, at 165–166.

93. Daniel Povinelli *et al.,* "Comprehension of seeing as a referential act in young children, but not juvenile chimpanzees," 17 *British Journal of Developmental Psychology* 37 (1999).

94. Krisztina Soproni *et al., supra* note 2, at 123.

95. *Id.* at 125.

96. *Id.*

97. Nicola J. Rooney *et al.,* "Do dogs respond to play signals given by humans?" 61 *Animal Behaviour* 715, 715 (2001); Maxeen Biben, "Comparative ontogeny of social behaviour in three South American canids, the maned wolf, crab-eating fox and bush dog: Implications for sociality," 31 *Animal Behaviour* 814, 822–824 (1983).

98. Nicola J. Rooney *et al., supra* note 97, at 716–718.

99. *Id.* at 718–721, 721.

100. Sz. Naderi *et al.,* "Cooperative interactions between blind persons and their dogs," 74 *Applications of Animal Behavioural Sciences* 59, 78–79 (2001).

101. *Id.* at 79.

102. Harold B. Weiss *et al.,* "Incidence of Dog Bite Injuries Treated in Emergency Departments," 279 *J. Amer. Med. Assoc.* 51, 53 (1998); *Statistical Abstract of the United States* (United States Census Bureau 1997); Mark Derr, *supra* note 32, at 194; Elizabeth C. Hirschman, "Consumers and Their Animal Companions," 20 *Journal of Consumer Re-*

search 616, 626 (1994); J. Karl Wise *et al.*, "Dog and cat ownership, 1991–1998," 204 *Journal of American Veterinary Medical Association* 1166 (1994).

103. Mark Derr, *supra* note 32, at 186–228.

104. Alan Beck and Aaron Katcher, *Between Pets and People—The Importance of Animal Companionship* 40–45 (2d ed. 1996); Mary Elizabeth Thurston, *The Lost History of the Canine Race: Our 15,000 Year-Love Affair with Dogs* 275 (1996); *The 1995 AAHA Report: A Study of the Companion Animals Veterinary Services* 84 (1995); Betty J. Carmack, "The effect of family members and functioning after the death of a pet," in *Pets and the Family* 149, 150 (Marvin B. Sussman, ed., 1985) (70 percent to 93 percent of American human companions view their companion animals as family members); Victoria L. Voith, "Attachment of people to companion animals," in 15 *Vet. Clin. North Am. [Small Anim. Prac.]* 289 (James Quackenbush and Victoria L. Voith, eds., 1985) (99 percent of 500 human companions surveyed); Ann Cain, "A study of pets in the family system," in *New Perspectives in Our Lives with Companion Animals* 5 (Alan Beck and Aaron Katcher, eds., 1983) (87 percent of 60 families surveyed); Thomas E. Catanzaro, "A study of the human-animal bond in military communities," in *The Pet Connection: Its Influence on Our Health and Quality of Life: Proceedings of the Minnesota-California Conferences on the Human Animal Bond* 341–347 (R. K. Anderson, B. L. Hart, and L. A. Hart 1984) (68 percent of 896 military families surveyed said that their companion animals were full family members; 30 percent said that their pets were close friends); Kenneth M. G. Keddie, "Pathological mourning after the death of a domestic pet," 131 *British Journal of Psychiatry* 21, 22 (1977).

105. Sandra B. Barker and Ralph T. Barker, "The human-canine bond: Closer than family ties?" 10(1) *Journal of Mental Health Counseling* 46 (January 1988).

106. *E.g.*, Harold A. Herzog and Shelley Gavin, "Common sense and the mental lives of animals: An empirical approach," in *Anthropormorphism, Anecdotes, and Animals* 237, 240–250 (Robert W. Mitchell *et al.*, eds., State University of New York Press 1997); Research and Marketing Services Department of Doyle Dane Bernbach (April 1983), reprinted at http://www.tufts.edu/vet/cfa/Surveys/welfint.html (visited August 15, 2001).

107. *Beuckner v. Hamel*, 886 S.W. 2d 368, 377–378 (Tex. App. 1993) (Anders, J., concurring).

108. Alan Beck and Aaron Katcher, *supra* note 104, at 41 (emphasis added); *id.* at 41–43 (companion animals are treated by families as young children); James Serpell, *In the Company of Animals* 63–70 (1986).

Chapter 8 Phoenix and Ake

1. I mailed three (October 15, 1999; March 5, 2000; March 9, 2001), and e-mailed three, one on April 23, 2001; one in April 2000, and one on October 5, 2001.

2. Personal communication from Louis Herman, dated April 23, 2001.

3. Personal communication from Louis Herman, dated October 5, 2001.

4. *Cetacean Behavior: Mechanisms and Functions* 432–433 (Louis M. Herman, ed., John Wiley & Sons 1980).

5. Richard C. Connor *et al.*, "The bottlenose dolphin," in *Cetacean Societies: Field Studies of Dolphins and Whales* 91, 92–93 (University of Chicago Press 2000).

6. Diana Reiss *et al.*, "Communicative and other cognitive characteristics of bottlenose dolphins," 1 *Trends in Cognitive Sciences* 140, 141 (July 1997).

7. When I refer to "dolphins," I mean Atlantic bottle-nosed dolphins.

8. P. J. Morgane *et al.,* "Evolutionary morphology of the dolphin brain," in *Dolphin Cognition and Behavior: A Comparative Approach* 5, 19 (Ronald J. Schusterman *et al.,* eds., Lawrence Erlbaum 1986).

9. Susan Greenfield, *The Private Life of the Mind* 61 (Penguin Press 2000). *See* Sue Taylor Parker and Michael L. McKinney, *Origins of Intelligence: The Evolution of Cognitive Development in Monkeys, Apes, and Humans* 326 (Johns Hopkins University Press 1999).

10. Sam H. Ridgeway, "Physiological observations on dolphins brains," in *Dolphin Cognition and Behavior: A Comparative Approach* 31, 35 (Ronald J. Schusterman *et al.,* eds., Lawrence Erlbaum Associates 1986).

11. Susan Greenfield, *supra* note 9, at 61.

12. Personal communication from Sue Taylor Parker, dated September 19, 2000.

13. Sue Taylor Parker and Michael L. McKinney, *supra* note 9, at 323, taken from A. Portmann, *A Zoologist Looks at Mankind* (J. Schaefer, trans., Columbia University Press 1990).

14. Susan Greenfield, supra *note* 9, at 65.

15. *But see* Kathleen Rita Gibson, "New Perspectives in instincts and intelligence: Brain size and the emergence of hierarchical mental construction skills," in *Intelligence and Languague in Monkeys and Apes* 97 (Sue Taylor Parker and Kathleen Rita Gibson, eds., Cambridge University Press 1999).

16. *E.g.,* Richard Byrne, *The Thinking Ape: Evolutionary Origins of Intelligence* 232 (Oxford University Press 1995).

17. *Id.* at 314, 324; Duane M. Rumbaugh, "Competence, cortex, and primate models," in *Development of the Prefrontal Cortex, Neurobiology, and Behavior* 117 (Norman A. Kasnegor *et al.,* eds., Paul H. Brooks Publishing Co., Inc. 1997).

18. Katerina Semendeferi, *et al.,* "The evolution of the frontal lobes: A volumetric analysis based on three-dimensional reconstructions of magnetic resonance scans of human and ape brains," 32 *Journal of Human Evolution* 375, 375 (1997).

19. *E.g.,* Susan Greenfield, *supra* note 9, at 146; Sue Taylor Parker and Michael L. McKinney, *supra* note 9, at 315.

20. Katerina Semendeferi, "The frontal lobes of the great apes with a focus on the gorilla and the orangutan," in *The Mentalities of Gorillas and Orangutans* 81, 83 (Sue Taylor Parker *et al.,* eds., Cambridge University Press 1999). Katerina Semendeferi *et al.,* "The volume of the cerebral hemispheres, frontal lobes, and cerebellum in living humans and apes using in vivo magnetic resonance morphometry," 24 *American Journal of Physical Anthropology, Supplement* 209–209 (1997); Katerina Semendeferi *et al., supra* note 18, at 375–388 (sample sizes were very small). *But see* Katerina Semendeferi *et al., supra* note 18, at 382–385; Dean Falk, "Evolution of the brain and cognition in hominids," *The Sixty-Second James Arthur Lecture on the Evolution of the Human Brain* (American Museum of Natural History, New York City 1992).

21. Sue Taylor Parker and Michael L. McKinney, *supra* note 9, at 323; 326; Katerina Semendeferi *et al., supra* note 20, at 380.

22. Katerina Semendeferi, "The frontal lobes of the great apes with a focus on the gorilla and the orangutan," in Sue Taylor Parker *et al.,* eds., *supra* note 20, at 70, 81; Sue Taylor Parker and Michael L. McKinney, *supra* note 9, at 326; Richard W. Byrne, "Cognitive

skills of the gorilla," in *Reaching Into Thought: The Minds of the Great Apes* 111, 125 (Cambridge University Press 1996).

23. Lori Marino, "A comparison of encephalization between odontocete cetaceans and anrthropoid primates," 51 *Brain, Behavior, and Evolution* 230 (1988).

24. Sam H. Ridgway, "Physiological observations on dolphin brains," in Ronald J. Schusterman *et al.*, eds., *supra* note 8, at 31, 32–33; P. J. Morgane *et al.*, supra note 8, at 22–23.

25. Diane Reiss *et al.*, *supra* note 6, at 141; Louis M. Herman and Palmer Morrel-Samuels, "Knowledge acquisition and asymmetry between language comprehension and production: Dolphins and apes as general models for animals," in *Interpretation and Explanation in the Study of Animal Behavior* 283, 300 (Marc Bekoff and Dale Jamieson, eds., Westview Press 1990); Sam H. Ridgway, *supra* note 10, at 45.

26. The manner is which Kea was trained is explained at Louis M. Herman, "Cognitive characteristics of dolphins," in Louis M. Herman, ed., *supra* note 4, at 363, 413–415. For a report on earlier work on communication with dolphins, *see generally*, John C. Lilly, *Communication Between Man and Dolphin: The Possibility of Talking with Other Species* (Crown Publishers 1978).

27. Louis M. Herman, *et al.*, "Comprehension of sentences by bottle-nosed dolphins," 16 *Cognition* 129, 134 (1984).

28. Personal communication from Kenneth Le Vasseur, dated March 27, 2001; Gavan Daws, "'Animal liberation' as crime: The Hawaii dolphin case," in *Ethics and Animals* 361, 361 (Harlan B. Miller and William H. Williams, eds., Humana Press 1983).

29. Louis M. Herman *et al.*, *supra* note 27, at 137.

30. Louis M. Herman, *supra* note 26 at 362.

31. *Id.* at 368.

32. Gavan Daws, "I Reach Beyond the Laboratory-Brain: Men, Dolphins, and Biography," in *Essaying Biography: A Celebration for Leon Edel* 167, 174 (Gloria Fromm, ed., University of Hawaii Press 1986).

33. Gavan Daws, *supra* note 28, at 362–371. *See generally*, Gavan Daws, *supra* note 32.

34. Personal communication from Louis Herman, dated April 23, 2001.

35. Louis M. Herman, *et al.*, *supra* note 27, at 135, note 2; Louis M. Herman, *supra* note 26, at 415; 1 *Oxford English Dictionary* 19 (2d ed., 1989) ("abducted," "abductor," "abduction," "abductee").

36. Louis M. Herman, *supra* note 26, at 415.

37. *Id.* For more on dolphin emotions, *see* Ronald J. Schusterman, "Pitching a fit," in *The Smile of a Dolphin: Remarkable Accounts of Animal Emotions* 106–107 (Marc Bekoff, ed., Discovery Books 2000); Toni Frohoff, "The dolphins' smile," in *id.* at 78–79.

38. Personal communication from Kenneth Le Vasseur, dated March 27, 2001.

39. Personal communication from Steven Sipman, dated July 30, 2001.

40. Melissa R. Shyan and Louis M. Herman, "Determinants of recognition of gestural signs in an artificial language by Atlantic bottle-nosed dolphins *(Tursiops truncatus)* and humans *(Homo sapiens)*," 101 *Journal of Comparative Psychology* 112, 112 (1987).

41. Louis M. Herman, *et al.*, *supra* note 27, at 137 note 3.

42. Personal communication from Steven M. Wise to Louis Herman, dated October 5, 2001.

43. Louis M. Herman *et al.*, *supra* note 27, at 137.

44. *Id.* at 137–139.

45. Herbert S. Terrace *et al.,* "Can an ape create a sentence?" 206 *Science* 892 (1979). *See* criticism of Terrace's "Project Nim," in Steven M. Wise, *Rattling the Cage: Toward Legal Rights for Animals* 172–174 (Perseus Publishing 2000).

46. Louis M. Herman, "Receptive competencies of language-trained animals," in 17 *Advances in the Study of Behavior* 1, 16 (J. S. Rosenblatt *et al.,* eds., Academic Press 1987).

47. Louis M. Herman and Robert K. Uyeyama, "The dolphin's grammatical competency: Comments on Kako (1999)," 27 *Animal Learning and Behavior* 18, 18–19 (1999); Louis M. Herman and Palmer Morrel-Samuels, "Knowledge acquisition and asymmetry between language comprehension and production: Dolphins and apes as general models for animals," in *Interpretation and Explanation in the Study of Animal Behavior*, 283, 283–301 (Marc Bekoff and Dale Jamieson, eds., Westview Press 1990); Louis M. Herman *et al., supra* note 13, at 209–210. *See* Irene Maxine Pepperberg, *The Alex Studies: Cognitive and Communicative Abilities of Grey Parrots* 126–127 (Harvard University Press 1999); Elizabeth Bates, "Comprehension and production in early language development," 58 *Monograph of the Society for Research in Child Development* 222, 223–238 (1993); E. Sue Savage-Rumbaugh *et al.,* "Language comprehension in ape and child," in *id.* at 16–23.

48. Louis M. Herman, *supra* note 46, at 1 and *id.,* quoting A. Paivo and I. Begg, *Psychology of Language* 25 (Prentice-Hall 1981); Louis M. Herman *et al., supra* note 27, at 130. *See* Edward Kako, "Elements of syntax in the systems of three language-trained animals," 27 *Animal Learning and Behavior* 1, 1–2 (1999).

49. Louis M. Herman *et al., supra* note 27, at 203–204.

50. Louis M. Herman, "The language of animal language research: Reply to Schusterman and Geisiner," 38 *The Psychological Record* 349, 351 (1988); Louis M. Herman *et al., supra* note 27, at 135.

51. Louis M. Herman *et al.,* supra note 27, at 135, 142–143. For a comparison between the way that humans and dolphins read gestural signs, *see* Melissa R. Shyan and Louis Herman, *supra* note 40.

52. Louis M. Herman, *supra* note 46, at 17.

53. Louis M. Herman *et al., supra* note 27, at 139–142.

54. Louis M. Herman, *supra* note 26, at 366.

55. Robert K. R. Thompson and Louis M. Herman, "Memory for lists of sounds by the bottle-nosed dolphin: Convergence of memory processes with human?" 195 *Science* 501, 501–502 (1977).

56. Adam A. Pack *et al.,* "Generalization of visual matching and delayed matching by a California sea lion *(Zalophus californianus),* 19 *Animal Learning and Behavior* 37, 39 (1991).

57. *Id.* at 380. *See* personal communication from Louis M. Herman, dated October 5, 2001.

58. Louis M. Herman, *supra* note 26, at 370–374.

59. Louis M. Herman, "What the dolphin knows, or might know, in its natural world," in *Dolphin Societies: Discoveries and Puzzles* 349, 353–354 (Karen Pryor and Kenneth S. Norris, eds., University of Chicago Press 2000); Louis Herman *et al.,* "Generalization of visual matching by a bottlenosed dolphin *(Tursiops truncatus):* Evidence for invariance of cognitive performance with visual or auditory materials," *Journal of Experimental Psychology: Animal Behavior Processes* 124, 134 (1989).

60. Eduardo Mercado III *et al.*, "Memory for recent actions in the bottlenosed dolphin *(Tursiops truncatus):* Repetition of arbitrary behaviors using an abstract rule," 26 *Animal Learning and Behavior* 210, 213–216 (1998).

61. *Id.* at 216.

62. Diagnostic Case Report, U.S. Geological Survey, Biological Resources Division, National Wildlife Health Center, Honolulu Field Station, Case #15400 (Final report, February 23, 2001).

63. Press release from The Dolphin Institute, dated May 24, 2001.

64. *Id.*

65. Press release from Animal Rights Hawaii, dated May 24, 2001.

66. I *Blackstone's Commentaries* *189 (18th London edition 1832).

67. Louis M. Herman, *supra* note 46, at 16.

68. *Id.*

69. Louis M. Herman *et al., supra* note 27, at 143–146; Louis M. Herman, *supra* note 46, at 33–34.

70. Louis M. Herman, *supra* note 46, at 50–51.

71. Louis M. Herman, "Cognition and language competencies of bottlenosed dolphins," in Ronald J. Schusterman *et al.*, eds., *supra* note 8, at 221, 230–231; Louis M. Herman *et al., supra* note 27, at 144, 147.

72. Louis M. Herman and Palmer Morrel-Samuels, *supra* note 25, at 298–299.

73. Louis M. Herman, *supra* note 46, at 17–18.

74. Louis M. Herman *et al., supra* note 27, at 188–189.

75. Louis M. Herman, *supra* note 71 at 230–231; Louis M. Herman *et al., supra* note 27, at 147, 162, 163, 165, 195.

76. Louis M. Herman, *et al., supra* note 27, at 197–198.

77. *Id.* at 190.

78. *Id.* at 198; Louis M. Herman, *supra* note 71, at 234–235.

79. Louis M. Herman *et al., supra* note 27, at 167. *See* Louis M. Herman, *supra* note 58, at 358.

80. Karen Pryor, "Reinforcement training as interspecies communication," in Ronald J. Schusterman *et al.*, eds., *supra* note 8, at 253, 255–256.

81. Louis M. Herman, *supra* note 46, at 51; Louis M. Herman *et al., supra* note 27, at 146–152.

82. Louis M. Herman *et al., supra* note 27, at 199.

83. Louis M. Herman and Palmer Morrel-Samuels, *supra* note 25, at 297.

84. Louis M. Herman *et al., supra* note 27, at 156–157.

85. Louis M. Herman and Palmer Morrel-Samuels, *supra* note 25, at 296; Louis M. Herman, *supra* note 46, at 19.

86. Louis M. Herman, *supra* note 46, at 27.

87. Louis M. Herman, *supra* note 71, at 233.

88. Louis M. Herman and Robert K. Uyeyama, *supra* note 47, at 19.

89. *Id.* at 19.

90. *Id.* at 19; Louis M. Herman *et al.*, "Responses to anomalous gestural sequences by a language-trained dolphin: Evidence for processing of semantic relations and syntactic information," 122 *Journal of Experimental Psychology: General* 184 (1993); Louis M. Her-

man, *supra* note 46, at 26; Louis M. Herman *et al., supra* note 27, at 197–198. Herman often cites Louis M. Herman *et al.,* "Processing of anomalous sentences by bottlenosed dolphins" (unpublished manuscript 1983) for his work on anomalous sentences. Herman did not produce this manuscript.

91. J. David Smith *et al.,* "The uncertain response in the bottlenosed dolphin (*Tursiops truncatus*)," 124 *Journal of Experimental Psychology: General* 391, 392–405 (1995).

92. Louis M. Herman, *supra* note 71, at 235; Louis M. Herman *et al., supra* note 27, at 184–188.

93. Louis M. Herman *et al., Journal of Experimental Psychology: General, supra* note 90, at 188–191.

94. Karen Pryor and Kenneth S. Norris, eds., *supra* note 59, at 356.

95. Louis M. Herman and Robert K. Uyeyama, *supra* note 47, at 21. Herman cites to C. G. Prince, "Conjunctive rule comprehension in a bottlenosed dolphin (unpublished master's thesis 1993). Herman did not produce this manuscript.

96. Louis M. Herman *et al., supra* note 27, at 177–178. Technically, the first are called semantic and context generalization; the second, displacement.

97. Louis M. Herman *et al., supra* note 27, at 158–159, 182.

98. Louis M. Herman, *supra* note 46, at 28; Louis M. Herman, *supra* note 71, at 238–239.

99. Louis M. Herman, *supra* note 46, at 27.

100. Louis M. Herman *et al., supra* note 27, at 178–182.

101. *Id.* at 183; Louis M. Herman, *supra* note 71, at 236.

102. Louis M. Herman, *supra* note 46, at 32.

103. Louis M. Herman *et al., supra* note 27, at 200.

104. *Id.* at 183–184.

105. Louis M. Herman, *supra* note 46, at 35.

106. Louis M. Herman and Palmer Morrel-Samuels, *supra* note 25, at 297.

107. Louis M. Herman, *supra* note 46, at 29; Louis M. Herman, *supra* note 71, at 239.

108. Louis M. Herman, *supra* note 46, at 21. *See* an earlier discussion, Louis M. Herman, *supra* note 71, at 231–233.

109. Louis M. Herman, *supra* note 46, at 21–22.

110. Louis M. Herman *et al., supra* note 27, at 190–193.

111. Louis M. Herman, *supra* note 71, at 240–241.

112. *Id.* at 233–234; Louis M. Herman, *supra* note 46, at 27.

113. Louis M. Herman, "In which Procrustean bed does the sea lion sleep tonight?" 39 *Psychological Record* 19, 42 (1989).

114. Louis M. Herman and Palmer Morrel-Samuels, *supra* note 25, at 299; Karen Pryor and Kenneth S. Norris, eds., *supra* note 59, at 357–358; Louis M. Herman, *supra* note 46, at 23; Louis M. Herman, *supra* note 71, at 239; Louis M. Herman *et al., supra* note 27, at 184.

115. Louis M. Herman *et al., supra* note 27, at 205.

116. Louis M. Herman *et al.,* "Representational and conceptual skills of dolphins," in *Language and Communication: Comparative Perspectives* 273 (Lawrence Erlbaum Associates 1993).

117. Louis M. Herman and Robert K. Uyeyama, *supra* note 47, at 20. Chance was one of six.

118. Louis M. Herman *et al.*, "Dolphins *(Tursiops truncatus)* comprehend the referential character of the human pointing gesture," 113 *Journal of Comparative Psychology* 347, 349–353 (1999).

119. *Id.* at 354–358.

120. *Id.* at 361–363.

121. Louis M. Herman, *supra* note 26, at 402.

122. Eugene Linden, *The Parrot's Lament and Other Tales of Animal Intrigue, Intelligence, and Ingenuity* 124 (Dutton 1999).

123. Diane Reiss *et al., supra* note 6, at 142.

124. Diane Reiss and Brenda McGowan, "Spontaneous vocal mimicry and production by bottlenose dolphins *(Tursiops truncatus):* Evidence for vocal learning," 107 *Journal of Comparative Psychology* 301, 301–311 (1993).

125. Louis M. Herman, *supra* note 26, at 405.

126. Diana Reiss, "The dolphin: An alien intelligence," in *First Contact: The Search for Extraterrestrial Intelligence* 31, 34, 38 (Ben Bova and Byron Preiss, eds., NAL Books 1990).

127. Peter L. Tyack, 289 *Science* 1310, 1310 (2000).

128. Diane Reiss, *et al., supra* note 6, at 142; Vincent M. Janik, "Whistle matching in wild bottlenose dolphins *(Tursiops truncatus),*" 289 *Science* 1355, 1355 (2000); Peter Tyack, "Deveolpment and social functions of signature whistles in bottlenose dolphins *(Tursiops truncatus),*" 8 *Bioacoustics* 21, 41–43 (1997).

129. Vincent M. Janik, *supra* note 128, at 1356–1357.

130. Douglas G. Richards *et al.*, "Vocal mimicry of computer-generated sounds and vocal labeling of objects by a bottle-nosed dolphin, *Tursiops truncatus,*" 98 *Journal of Comparative Psychology* 10, 13–14 (1984).

131. Louis M. Herman, *supra* note 46, at 30; Douglas G. Richards *et al., supra* note 130, at 16.

132. Douglas G. Richards *et al., supra* note 130, at 17–20.

133. Louis M. Herman, "Seeing through sound: Dolphins *(Tursiops truncatus)* perceive the spatial structure of objects through echolocation," 112 *Journal of Comparative Psychology* 292, 304 (1998).

134. Katherine A. Loveland, "Self-recognition in the bottlenose dolphin: Ecological considerations," 4 *Consciousness and Cognition* 254, 255–256 (1995); Diana Reiss, "Self-view television as a test of self-awareness: Only in the eye of the beholder?" 4 *Consciousness and Cognition* 235, 237 (1995).

135. Kenneth Marten and Suchi Psarakos, "Marten and Psarakos commentary response," 4 *Consciousness and Cognition* 258, 264 (1995).

136. F. Delfourab and K. Marten, "Mirror image processing in three marine mammal species: Killer whales *(Orcinus orca),* false killer whales *(Pseudorca crassidens)* and California sea lions *(Zalophus californianun),*" (2001) (in press).

137. Lori Marino *et al.*, "Mirror self-recognition in bottlenose dolphins: Implications for comparative investigations of highly dissimilar species," in *Self-Awareness in Animals and Humans: Developmental Perspectives* 380, 387–390 (Sue Taylor Parker *et al.*, eds., Cambridge University Press 1994).

138. Kenneth Marten and Suchi Psarakos, "Evidence of self-awareness in the bottlenose dolphin *(Tursiops truncatus),*" in Sue Taylor Parker *et al.*, eds., *supra* note 137, at 361, 362, 366, 367, 373, 374.

139. Kenneth Marten and Suchi Psarakos, "Using self-view television to distinguish between self-examination and social behavior in the bottlenose dolphin *(Tursiops truncatus),*" 4 *Consciousness and Cognition* 205, 206–213, 216–217 (1995); Kenneth Marten and Suchi Psarakos, *supra* note 138, at 372.

140. Kenneth Marten and Suchi Psarakos, *supra* note 139, at 218–219, 222; Kenneth Marten and Suchi Psarakos, *supra* note 138, at 374, 378.

141. Diana Reiss and Lori Marino, "Mirror self-recognition in the bottlenose dolphin: A case of cognitive convergence," 98(10) *Proceedings of the National Academy of Sciences* 5937, 5942 (May 8, 2001). Daniel Povinelli and Gordon Gallup have denied that this dolphin MSR data is definitive, Philip Yam, "The Flipper effect," 285(1) *Scientific American* 29 (July 2001).

142. Diana Reiss and Lori Marino, *supra* note 141, at 5942.

143. Alain Tschudin *et al.,* "Comprehension of signs by dolphins *(Tursiops truncatus),*" 115(1) *Journal of Comparative Psychology* 100, 101, 102, 104 (2001).

144. Peter L Tyack, *supra* note 127, at 1311.

145. Rachel Smolker, *To Touch a Dolphin* 107–117 (Doubleday 2001).

146. Peter L Tyack, *supra* note 127, at 1311.

147. Richard C. Connor *et al., supra* note 5, at 91, 103–104, 108, 109, 118.

148. *Id.* at 111–113, 124. *See* Richard C. Connor *et al.,* "Complex soxial structure, alliance stability and mating access in a bottlenose dolphin 'super-alliance,'" 268 *Proceedings of the Royal Society of London* 263, 263–266 (2001) (Indian bottlenosed dolphins).

149. Bernd Wursig, "The question of dolphin awareness approached through studies in nature," 5 *Cetus* 1 (1985).

150. Karen Pryor *et al.,* "A dolphin-human fishing cooperative in Brazil," 6 *Marine Mammal Science* 77 (1990).

151. Denise L. Herzing amd Thomas I. White, "Dolphins and personhood," in *Etica & Animali* 64, 81–82 (September 1998), citing Susan Shane, *The Bottlenose Dolphin in the Wild* 5.

152. Richard C. Connor *et al., supra* note 5, at 104.

153. *Id.* at 115–117.

154. Rachel Smolker, *supra* note 145, at 256, 263.

155. Conversation with Ronald J. Schusterman, November 20, 2000. *See* Ronald J. Schusterman and Robert Gisiner, "Please parse the sentence: Animal cognition in the Proscrustean bed of linguistics," 39 *The Psychological Record* 3 (1989); Ronald J. Schusterman and Robert Gisiner, "Artificial language comprehension in dolphins and sea lions: The essential cognitive skills," 38 *The Psychological Record* 311 (1998). For Herman's responses, *see* Louis M. Herman, *supra* note 50, and Louis M. Herman, *supra* note 113.

156. Ronald J. Schusterman and Robert Gisiner ("Artificial language comprehension"), *supra* note 155, at 319.

157. Ronald Gisiner and Ronald J. Schusterman, "Sequence, syntax, and semantics: Responses of a language-trained sea lion *(Zalophus californianus)* to novel sign combinations," 106 *Journal of Comparative Psychology* 78, 78–86 (1992); Ronald J. Schusterman and Ronald Gisiner, "Pinnipeds, porpoises, and parsimony: Animal language research viewed from a bottom-up perspective," in Robert W. Mitchell *et al.,* eds., *Anthropomorphism, Anecdotes, and Animals* 370–382 (State University of New York Press 1997); Ronald J. Schusterman and Kathy Krieger, "Artifical language comprehension and size transposi-

tion by a California sea lion *(Zalophus californianus),"* 100 *Journal of Comparative Psychology* 348, 348–349 (1986); Ronald J. Schusterman and Kathy Krieger, "California sea lions are capable of semantic comprehension," 34 *The Psychological Record* 3–23 (1984).

158. Ronald Gisiner and Ronald J. Schusterman, *supra* note 157, at 86. See Louis M. Herman, *supra* note 46, at 48–49.

159. Adam A. Pack *et al., supra* note 56, at 47.

160. Ronald J. Schusterman *et al.,* "How animals classify friends and foes," 9 *Current Trends in Psychological Science* 1, 1 (February 2000).

161. *Id.* at 6.

162. Harold A. Herzog and Shelley Gavin, "Common sense and the mental lives of animals: An empirical approach," in Robert W. Mitchell *et al.,* eds., *supra* note 157, at 237, 240–250.

Chapter 9 Echo

1. Until her 1997 killing by the Kenyan Wildlife Service, the matriarch, Tuskless, held that distinction, leading her family through and around Moss's camp and nearby tourist lodges, and being featured in more than a hundred documentary films, Cynthia Moss, *Elephant Memories: Thirteen Years in the Life of an Elephant Family* (with a new *Afterword*) 335 (University of Chicago Press 2000) (1988).

2. Karl Groning and Maryin Saller, *Elephants: A Cultural and Natural History* 58 (Koneman, undated).

3. Personal communication from Cynthia Moss, dated August 3, 2001.

4. Jeheskel Shoshani and Pascal Tassy, "Classifying elephants," in *Elephants: Majestic Creatures of the Wild* 26–27 (Jeheskel Shoshani, ed., Checkmark Books 2000); Joyce H. Poole, "An exploration of a commonality between ourselves and elephants," in *Etica and Animali* (special issue devoted to nonhuman personhood) 85, 94 (September 1998); Joyce Poole, *Coming of Age with Elephants* 133 (Hodder and Stoughton 1996).

5. Personal communication from "Mr. Darwin," dated May 1, 2001.

6. Conversation with Cynthia Moss at Amboseli, May 31, 2001; Cynthia Moss, *supra* note 1, at 219.

7. Personal communication from Cynthia Moss, dated August 3, 2001.

8. Cynthia Moss, *supra* note 1, at 36.

9. Personal communication from Cynthia Moss, dated August 3, 2001.

10. Personal communication from Iain Douglas-Hamilton, dated July 16, 2001.

11. Katy Payne, *Silent Thunder: In the Presence of Elephants* 45, 54 (Simon & Schuster 1998); Joyce H. Poole, *supra* note 4, at 88; Cynthia Moss and Martyn Colbeck, *Echo of the Elephants: The Story of an Elephant Family* 40 (William Morrow and Co. 1992); Cynthia Moss, *supra* note 1, at 38, 130, 131.

12. Cynthia Moss, *supra* note 1, at 35. *See* Joyce H. Poole, *supra* note 4, at 88, 89; Joyce Poole, *supra* note 4, at 29.

13. Cynthia Moss, *supra* note 1, at 38. By 1999, the TCs and TDs seemed no longer part of this bond group, *id.* at 327.

14. Joyce H. Poole, *supra* note 4, at 88; Cynthia Moss and Martyn Colbeck, *supra* note 11, at 40; Cynthia Moss, *supra* note 1, at 76.

15. Cynthia Moss, *supra* note 1, at 126, 132.

16. Personal communication from Cynthia Moss, dated August 3, 2001.

17. Katy Payne, *supra* note 11, at 55. Cynthia Moss suspects that bond groups may be genetic as well, personal communication from Cynthia Moss, dated August 3, 2001.

18. Cynthia Moss and Martyn Colbeck, *supra* note 11, at 47.

19. Joyce H. Poole, supra note 4, at 87; Cynthia Moss, *supra* note 1, at 35.

20. Cynthia Moss and Martyn Colbeck, *supra* note 11, at 149.

21. Katy Payne, *supra* note 11, at 100.

22. *Id.* at 196–197.

23. Personal communication from Cynthia Moss, dated August 3, 2001; conversation with Cynthia Moss at Amboseli, May 31, 2001.

24. Cynthia Moss and Martyn Colbeck, *supra* note 11, at 25.

25. *Id.* at 44.

26. Conversation with Cynthia Moss, May 31, 2001; Cynthia Moss and Martyn Colbeck, *supra* note 11, at 32.

27. Joyce H. Poole, *supra* note 4, at 91; Katy Payne, *supra* note 11, at 54; Cynthia Moss, *supra* note 1, at 128. One may see examples of these greetings on video, "Echo of the Elephants: The Next Generation," at 33:45; "Echo of the Elephants," at 13:45.

28. Joyce H. Poole, *supra* note 4, at 91.

29. Charles Darwin, *The Expression of the Emotions in Man and Animals* 168 (3d ed. Oxford University Press 1998).

30. Karl Groning and Maryin Saller, *supra* note 2, at 435; Jeffrey Moussaieff Masson and Susan McCarthy, *When Elephants Weep: The Emotional Lives of Animals* 106–108 (Delacorte Press 1995); Douglas H. Chadwick, *The Fate of the Elephant* 327 (Sierra Club Books 1992).

31. Iain and Oria Douglas-Hamilton, *Among the Elephants* 231 (Viking Press 1975).

32. Joyce H. Poole, *supra* note 4, at 90.

33. *Id.* at 90–91.

34. Cynthia Moss and Martyn Colbeck, *supra* note 11, at 56.

35. Cynthia Moss, *supra* note 1, at 125.

36. Douglas H. Chadwick, *supra* note 30, at 67. For a photograph of a group of marksmen in midannihilation of an elephant family, *see* Douglas Chadwick, "Elephants: Out of time, out of space," *National Geographic* 11, 44–45 (May 1991).

37. Anthony J. Hall-Martin, "The question of culling," in Jeheskel Shoshani, ed., *supra* note 4, at 194, 194–199, 201; Katy Payne, *supra* note 11, at 69–70, 199–200; Joyce Poole, *supra* note 4, at 158, Douglas H. Chadwick, *supra* note 30, at 430–432; Cynthia Moss, *supra* note 1, at 315–316. Cynthia Moss finds no sound evidence that African elephants ever exceed their carrying capacity; their numbers naturally rise and fall as resources become more or less available, Cynthia Moss and Martyn Colbeck, *supra* note 11, at 142–144.

38. Ian M. Redmond, "Alternatives to culling," in Jeheskel Shoshani, ed., *supra* note 4, at 200; Katy Payne, *supra* note 11, at 68, 197.

39. Joyce H. Poole, *supra* note 4, at 92.

40. Conversation with Collins Ajouk, Ngong Hills, Nairobi, May 30, 2001.

41. Karl Groning and Maryin Saller, *supra* note 2, at 58.

42. Francis Bacon, "On revenge," in *Essays* 15 (Prometheus Books 1995).

43. Joyce H. Poole, *supra* note 4, at 92.

44. Douglas H. Chadwick, *supra* note 30, at 326.

45. Karen McComb *et al.,* "Matriarchs as repositories of social knowledge in African elephants," 292 *Science* 491, 492, 493 (April 20, 2001). *See* Karen McComb *et al.,* " Unusually extensive networks of vocal recognition in African Elephants," 59 *Animal Behavior* 1103, 1107–1108 (2000).

46. Bernard Rensch, "The intelligence of elephants," 196 *Scientific American* 44, 48 (1956).

47. Recent data strongly suggest that two species of African elephants exist, *Loxodonta africana,* the savannah elephant, and the smaller forest elephant, the proposed name being *Loxodonta cyclotis,* Alfred L. Roca *et al.,* "Genetic evidence for two species of elephant in Africa," 293 *Science* 1473 (2001).

48. Personal communication from Cynthia Moss, dated August 3, 2001.

49. Jeheskel Shoshani, "Anatomy and physiology," in Jeheskel Shoshani, ed., *supra* note 4, at 200; conversation with Cynthia Moss at Amboseli, May 31, 2001.

50. Hal Markowitz, *Behavioral Enrichment in the Zoo* 89 (Van Rostrand Reinhold Co. 1982); Hal Markowitz *et al.,* "Do elephants ever forget?" 3 *Journal of Applied Behavior Analysis* 333, 334 (1975).

51. David Gucwa and James Ehmann, *To Whom It May Concern: An Investigation of the Art of Elephants* 127 (W. W. Norton 1985); Victor J. Stevens, "Basic operant research in the zoo," in *Behavior of Captive Wild Animals* 209, 212–214 (Hal Markowitz and Victor J. Stevens, eds., Nelson-Hall 1978).

52. Charles Winton Hyatt, "Discrimination learning in the African elephant," unpublished master's thesis for the faculty of the division of graduate studies at the Georgia Institute of Technology 13 (May 1991). I am grateful to Dr. Tara Stoinski for locating and providing me with a copy of Hyatt's thesis. *See* Jesheskel Shoshani and John F. Eisenberg, "Intelligence and survival," in Jeheskel Shoshani, ed., *supra* note 4, at 134, 135; Joyce H. Poole, *supra* note 4, at 95; Douglas H. Chadwick, *supra* note 30, at 280. At the National Zoo, Ben Beck confirms that Hyatt studied elephants both there and at Zoo Atlanta; ZooAtlanta's director, Terry Maple, says that Hyatt worked as his student. Maple recalls the results being mixed and says they obtained good data with only one elephant, personal communication from Ben Beck, dated May 13, 2001; personal communication from Terry Maple, dated May 8, 2001.

53. Charles Winton Hyatt, *supra* note 52, at 19.

54. Douglas H. Chadwick, *supra* note 30, at 280–281. My copy of Hyatt's thesis says that "the human experimenters relied heavily on written notes to keep the S+/S- straight," Charles Winton Hyatt, *supra* note 52, at 39.

55. Charles Winton Hyatt, *supra* note 52, at 38.

56. *Id.* at 40.

57. Robert H. I. Dale *et al.,* "Spatial memory abilities of captive female African elephants *(Loxodonta africana)* (in preparation); Robert H. I. Dale *et al.,* "Long-term retention of a short-term memory task by five female African elephants *(Loxodonta africana),*" *Proceedings of the 14th Annual Elephant Managers Conference* 12, 12–13 (October 16–19, 1993, Marine World Africa USA, Vallejo, California); Robert H. I. Dale *at al.,* "Preliminary studies of the spatial memory abilities of captive African elephants *(Loxidonta africana),*" *The 13th International Elephant Workshop* 55, 55–56 (1993, ZooAtlanta, Atlanta, Georgia).

58. Robert H. I. Dale *et al.* (1993 *Proceedings*), *supra* note 57, at 13–14.

59. Robert H. I. Dale *et al.*, "Both African and Asian elephants exhibit spatial memory with olfactory cues controlled," *Poster presented at the 10th Annual Convention of the American Psychological Society*, Washington, D.C., May 21–24, 1998.

60. Robert H. I. Dale *et al.*, "Situational correlates of a head-shake behavior by captive female African elephants *(Loxodonta africana)*," *Poster presented at the Annual Meeting of the Animal Behavior Society*, Atlanta, Georgia, August 5–9, 2000; Robert H. I. Dale *et al.*, "Situational correlates of visual displays of female African elephants *(Loxidonta africana)* in a controlled situation," *Poster Presented at the Elephant Manager's Association at the AAZK/EME/AZH National Conference,* Indianapolis, Indiana, 1998.

61. Daniel J. Povinelli, "Failure to find self-recognition in Asian elephants *(Elephas maximus)* in contrast to their use of mirror cues to discover hidden food," 103 *Journal of Comparative Psychology* 122 (1989).

62. Lesley J. Rogers, *Minds of Their Own: Thinking and Awareness in Animals* 29–30 (Westview Press 1998).

63. Personal communication from Joyce Poole, dated April 27, 2001; Patricia Simonet *et al.*, "Social and cognitive factors in Asian elephant mirror behavior and self-recognition," at 1–3 (unpublished paper, 2000) (presented to the Animal Social Complexity and Intelligence Conference [Chicago Academy of Sciences August, 2000)] [Abstract #25]).

64. Plutarch, "Whether land or sea animals are cleverer," in XII *Moralia* 319, 373–374 (Harold Cherness, trans., Harvard University Press 1984). Thanks to Christopher Jones for locating this citation and Martha Nussbaum for referring me to Professor Jones.

65. Joyce Poole, *supra* note 4, at 158.

66. Personal communication from Joyce Poole, dated April 27, 2001; Joyce H. Poole, *supra* note 4, at 88; Joyce Poole, *supra* note 4, at 154.

67. Cynthia Moss and Martyn Colbeck, *supra* note 11, at 61; Cynthia Moss, *supra* note 1, at 270.

68. Conversation with Cynthia Moss at Amboseli, May 31, 2001. *See* Douglas Chadwick, *supra* note 36, at 40–41.

69. Katy Payne, *supra* note 11, at 64.

70. Cynthia Moss and Martyn Colbeck, *supra* note 11, at 60–61; Katy Payne, *supra* note 11, at 64 (lion kill); Joyce Poole, "When bonds are broken," in *The Smile of a Dolphin: Remarkable Accounts of Animal Emotions* 142, 142 (Marc Bekoff, ed., Discovery Books 2000); Cynthia Moss, *supra* note 1, at 74, 270; Iain and Oria Douglas-Hamilton, *supra* note 31, at 233–239.

71. Douglas Chadwick, *supra* note 36, at 40.

72. Cynthia Moss and Martyn Colbeck, *supra* note 11, at 60; Cynthia Moss, *supra* note 1, at 270.

73. Conversation with Cynthia Moss at Amboseli, May 31, 2001.

74. Cynthia Moss and Martyn Colbeck, *supra* note 11, at 60.

75. Cynthia Moss, *supra* note 1, at 336.

76. Conversation with Cynthia Moss at Amboseli, May 31, 2001.

77. Marc Hauser, *Wild Minds: What Animals Really Think* 225, 226, 227 (Henry Holt 2000).

78. Douglas H. Chadwick, *supra* note 36, at 11, 12, 15, 17. *See* Benjamin B. Beck, *Animal Tool Behavior: The Use and Manufacture of Tools by Animals* 33–35 (Garland STPM Press 1980).

79. Jesheskel Shoshani and John F. Eisenberg, "Intelligence and survival," in Jeheskel Shoshani, ed., *supra* note 4, at 134, 136.

80. Joyce Poole, *supra* note 4, at 132.

81. Bernard Rensch, *supra* note 46, at 47.

82. *Id.* at 47.

83. *Id.* at 48.

84. *Id.* at 48–49.

85. Suzanne Chevalier-Skolnikoff and Jo Liska, "Tool use by wild and captive elephants," 46 *Animal Behavior* 209, 209 (1993).

86. *Id.* at 210–211 (1993).

87. *Id.* at 210.

88. *Id.* at 212, 213.

89. *Id.* at 214, 215.

90. Katy Payne, *supra* note 11, at 252.

91. James A. Gordan, "Elephants do think," 20 *African Wild Life* 75, 75–79 (1966).

92. Douglas H. Chadwick, *supra* note 30, at 69.

93. Katy Payne, *supra* note 11, at 97; Joyce H. Poole, *supra* note 4, at 93. *See* http://www.coolmath.com/decibels1.htm (visited January 8, 2002).

94. Personal communication from Joyce Poole, dated August 11, 2001; conversation with Joyce Poole at Nairobi, May 27, 2001; Joyce H. Poole, *supra* note 4, at 93; Joyce Poole, *supra* note 4, at 123, 136–137. Some of these calls can be heard on video, "Echo of the Elephants: The Next Generation," at 8:10, 11:00, 26:30, 33:45, 40:30, and 48:20; "Echo of the Elephants," at 10:24, 11:49, 13:45, and 45:45.

95. Personal communication from Joyce Poole, dated August 11, 2001; Joyce H. Poole, *supra* note 4, at 93; Joyce Poole, *supra* note 3, at 136.

96. Joyce H. Poole, *supra* note 4, at 97; Joyce Poole, *supra* note 4, at 132, 149–150. Douglas Chadwick tells a similar story of Joshua and the flip-flop, complete with wildebeest bone head-bonking, that he observed, in Douglas H. Chadwick, *supra* note 30, at 59–60.

97. Douglas Chadwick, *supra* note 36, at 25.

98. Cynthia Moss, *supra* note 1, at 163.

99. Cynthia Moss and Martyn Colbeck, *supra* note 11, at 46–47, 53, 59, 63, 80, 81, 162, 164; Cynthia Moss, *supra* note 1, at 75, 85–86, 142–143, 165, 168–169, 171, 322–323. *See* Joyce Poole *supra* note 4, at 148.

100. Joyce H. Poole, *supra* note 4, at 106.

101. Douglas H. Chadwick, *supra* note 36, at 15.

102. *Id.* at 12.

103. Joyce H. Poole, *supra* note 4, at 105; "Echo of the Elephants," at 9:03.

104. Joyce H. Poole, *supra* note 4, at 105; Joyce Poole *supra* note 4, at 139.

105. Robert H. I. Dale, "When Broadway met the Big Top: The Story of the Circus Polka," *Presentation Given at the 17th Annual Elephant Managers Workshop,* Jacksonville Zoological Gardens, January 1997.

106. Douglas H. Chadwick, *supra* note 30, at 475.

107. Richard Leaky and Virginia Morrell, *Wildlife Wars: My Fight to Save Africa's Natural Treasures* 154–156 (St. Martin's Press 2001).

108. Colin Tudge, "Does ivory burn?" 413 *Nature* 353, 353 (2001).

Chapter 10 Chantek

1. Personal communication from Daniel Shillito (undated in 2000). He's not the only orangutan expert who thinks this, personal communication from Anne Russon, dated January 28, 2001.

2. Personal observations made on February 16, 2000, at Zoo Atlanta, Atlanta, Georgia.

3. H. Lyn Miles *et al.,* "Simon says: The development of imitation in an enculturated orangutan," in *Reaching Into Thought: The Minds of the Great Apes* 282 (Anne E. Russon *et al.*, eds., Cambridge University Press 1996).

4. H. Lyn White Miles, "ME CHANTEK: The development of self-awareness in a signing orangutan," in *Self-awareness in Animals and Humans: Developmental Perspectives* 254, 255–257 (Sue Taylor Parker, *et al.,* eds., Cambridge University Press 1994); H. Lyn White Miles, "The cognitive foundations for reference in a signing orangutan, in *Intelligence and Language in Monkeys and Apes* 511, 512–513 (Sue Taylor Parker and Kathleen Rita Gibson, eds., Cambridge University Press 1990); H. Lyn Miles, "Apes and language: The search for communicative competence," in *Language in Primates: Perspectives and Implications* 43, 47 (Judith de Luce and Hugh T. Wilder, eds., Springer-Verlag 1983).

5. Anne E. Russon, *Orangutans: Wizards of the Rain Forest* 25 (Firefly Books 2000); Birute M. F. Galdikas, *Reflections of Eden: My Years With the Orangutans of Borneo* 43, 93 (Little, Brown 1995).

6. Anne E. Russon, *supra* note 5, at 40.

7. David R. Begun, "Hominid family values: Morphological and molecular data on relations among the great apes and humans," in *The Mentalities of Gorillas and Orangutans* 11–12 (Sue Taylor Parker *et al.,* eds., Cambridge University Press 1999).

8. Morris Goodman *et al.,* "Primate phylogeny and classification elucidated at the molecular level," in *Evolutionary Theory and Practice: Modern Perspectives* 193, 207 (S. P. Wasser, ed., Kluwer Academic Publishers 1999).

9. David R. Begun, *supra* note 7, at 13–25. *E.g.*, Morris Goodman *et al.,* "Toward a phylogenetic classification of primates based on DNA evidence complemented by fossil evidence," 9(3) *Molecular Phylogenetics and Evolution* 585, 594 (1998); Richard Wrangham and Dale Peterson, *Demonic Males: Apes and the Origins of Human Violence* 42 (Mariner Books 1996); W. J. Bailey *et al.,* "Re-examination of the African hominid trichotomy with additional sequences from the primate beta-globin gene cluster," 1 *Molecular Phylogenetics and Evolution* 97 (1992).

10. Birute M. F. Galdikas, *supra* note 5, at 131.

11. Anne E. Russon, *supra* note 5, at 35. In captivity, orangutans mature much earlier, personal communication from Robert W. Shumaker (undated in 2001).

12. Anne E. Russon, *supra* note 5, at 35; Cheryl Knott, "Orangutans in the Wild," 194 *National Geographic* 30, 54 (August 1998).

13. Anne E. Russon, *supra* note 5, at 35.

14. Personal communication from Anne Russon, dated January 28, 2001; Anne E. Russon *et al.,* "Seeing with the mind's eye: Eye-covering play in orangutans and Japanese macaques?" in *Pretending and Imagination in Animals and Children* (Robert W. Mitchell, ed., Cambridge University Press 2002).

15. Personal communication from Anne Russon, dated January 28, 2001; R. K. Davenport, "The orangutan in Sabah," 5 *Folia Primatologica* 247 (1967).

16. Gisela Kaplan and Lesley J. Rogers, *The Orangutans: Their Evolution, Behavior, and Future* 37 (Perseus Publishing 2000); John R. MacKinnon, *The Ape Within Us* (Collins 1978).

17. Personal communication from Carel van Schaik, dated January 12, 2001; Birute F. M. Galdikas and P. Vasey, "Why are orangutans so smart? Ecological and social hypothesis," in *Social Processes and Mental Abilities in Nonhuman Primates* 183, (F. D. Burton, ed., Edward Mellen Press 1992).

18. Elizabeth A. Fox *et al.*, "Intelligent tool use in wild Sumatran orangutans," in Sue Taylor Parker *et al.*, eds, *supra* note 7, at 99, 99; Birute M. F. Galdikas, *supra* note 5, at 119.

19. Personal communication from Elizabeth Fox, dated March 22, 2001.

20. *Id.*

21. *Id.*

22. *Id.;* Personal communication from Carel van Schaik, dated January 12, 2001.

23. Carel P. van Schaik *et al.*, "The conditions for tool use in primates: Implications for the evolution of material culture," 36 *Journal of Human Evolution* 719, (1999); Carel P. van Schaik and Cheryl Knott, "Orangutan cultures," American Association of Physical Anthropology, abstracts, at 223 (1999). Their tool use within discrete populations may be the best evidence that wild orangutans, like wild gorillas and chimpanzees, have cultures.

24. Michael Tomasello and Josep Call, *Primate Cognition* 88 (Oxford University Press 1997); C. E. Parker, "Responsiveness, manipulation, and implementation behavior in chimpanzees, gorillas, and orang-utans," in *Proceedings of Second International Congress on Primatology* 160, (C. R. Carpenter, ed., Karger 1969).

25. Personal communication from Anne Russon, dated January 28, 2001.

26. Jurgen Lethmate, "Tool-using skills of orang-utans," 11 *Journal of Human Evolution* 49, (1982).

27. Personal communication from Anne E. Russon, dated January 28, 2001; Anne E. Russon, *supra* note 5, at 59.

28. *Id.* at 62–63.

29. *Id.* at 59.

30. Kathy D. Schick *et al.*, "Continuing investigations into the stone tool-making and tool-using capacities of a bonobo *(pan paniscus)*," 26 *Review of Archeaeological Science* 821, 821–822 (1999).

31. Personal communication from Carel van Schaik, dated January 12, 2001; R. V. Wright, "Imitative learning of a flaked tool technology: The case of an orangutan," 8 *Mankind* 296 (1972).

32. Bernd Heinrich, "Testing insight in ravens," in *The Evolution of Cognition* 289, 289, 300–301 (Cecelia Heyes and Ludwig Huber, eds., MIT Press 2000); Sandra T. de Blois *et al.*, "Object permanence in orangutans *(Pongo pygmaeus)* and squirrel monkeys *(Saimiri sciureus)*," 112 *Journal of Comparative Psychology* 137, 148 (1998); Michael Tomasello and Josep Call, *supra* note 24, at 68–69.

33. Personal communication from Carel van Schaik, dated January 12, 2001; Gisela Kaplan and Lesley J. Rogers, *supra* note 16, at 11; Elizabeth A. Fox *et al., supra* note 18, at 101–111; Michael Tomasello and Josep Call, *supra* note 24, at 89; Suzanne Chevalier-

Skolnikoff, "Sensorimotor development in orangutans and other primates," 12 *Journal of Human Evolution* 545 (1983).

34. Personal communication from Elizabeth Fox, dated March 22, 2001.

35. Carel P. van Schaik *et al.*, "The conditions for tool use in primates: Implications for the evolution of material culture," 36 *Journal of Human Evolution* 719 (1999).

36. Anne E. Russon *et al., supra* note 14.

37. H. Lyn White Miles (Sue Taylor Parker and Kathleen Rita Gibson, eds.), *supra* note 4, at 530.

38. H. Lyn White Miles (Sue Taylor Parker *et al.*, eds.), *supra* note 4, at 263; H. Lyn White Miles (Sue Taylor Parker and Kathleen Rita Gibson, eds.), *supra* note 4, at 533.

39. Jurgen Lethmate, *supra* note 26.

40. Gisela Kaplan and Lesley J. Rogers, *supra* note 16, at 82.

41. Josep Call, "Object permanence in orangutans *(Pongo pygmaeus),* chimpanzees *(Pan troglodytes),* and children *(Homo sapiens),*" 115(2) *Journal of Comparative Psychology* 159, 160–170 (2001).

42. Irene Maxine Pepperberg, *The Alex Studies: Cognitive and Communicative Abilities of Grey Parrots* 112–113 (Harvard University Press 1999).

43. Josep Call, "Estimating and operating on discrete quantities in orangutans *(Pongo pygmaeus),*" 114 *Journal of Comparative Psychology* 136 (2000). *See also,* Josep Call and Phillippe Rochat, "Perceptual Strategies in the *Estimation of Physical Quantities by Orangutans (Pongo pygmaeus),*" 111 *Journal of Comparative Psychology* 315 (1997).

44. Josep Call, *supra* note 43.

45. Michael Cole and Sheila R. Cole, *The Development of Children* 476 (4th ed., Worth Publishers 2001).

46. *Id.* at 338–339, 447–448.

47. Josep Call and Philippe Rochat, "Liquid conservation in orangutans *(Pongo pygmaeus)* and humans *(Homo sapiens):* Individual differences and perceptual strategies," 110 *Journal of Comparative Psychology* 219, 230–231 (1996).

48. *Id.*

49. *Id.* at 231.

50. Josep Call and Philippe Rochat, *supra* note 43, at 315, 327–328.

51. Gordon G. Gallup and Daniel J. Shillito, Review of Sue Taylor Parker *et al.*, eds., *supra* note 7, in 75 *The Quarterly Review of Biology* 352, 352 (2000); Karyl B. Swartz *et al.*, "Comparative aspects of mirror self-recognition in great apes," in Sue Taylor Parker *et al.*, eds., *supra* note 4, at 283, 287, 288; Ethel Tobach *et al.*, "Viewing of self and nonself images in a group of captive orangutans *(Pongo pygmaeus abelii),*" 84 *Perceptual and Motor Skills* 355 (1997); Michael Tomasello and Josep Call, *supra* note 24, at 331; Karyl B. Swartz and Sian Evans, "Anthropomorphism, anecdotes, and mirrors," in *Anthropomorphism, Anecdotes, and Animals* 303–304 (Robert W. Mitchell *et al.*, State University of New York Press 1997); H. Lyn White Miles (Sue Taylor Parker *et al.*, eds.), *supra* note 4, at 257; Adrian Kortlandt, "Primates in the looking glass: A historical note," 37 *Animal Behavior Society Newsletter* 11, 11 (1991).

52. Personal communication from Daniel Shillito (undated in 2000).

53. H. Lyn White Miles (Sue Taylor Parker *et al.*, eds.), *supra* note 4, at 262, 263.

54. H. Lyn White Miles (Sue Taylor Parker *et al.*, eds.), *supra* note 4, at 264, 266–267.

55. Josep Call and Michael Tomasello, "The effect of humans on the cognitive development of apes," in Anne E. Russon, *et al.,* eds., *supra* note 3, at 370, 382–383.

56. Gisela Kaplan and Lesley J. Rogers, *supra* note 16, at 107.

57. H. Lyn White Miles (Sue Taylor Parker and Kathleen Rita Gibson, eds.), *supra* note 4, at 525.

58. Personal communication from Lyn Miles, dated October 26, 2000; H. Lyn White Miles (Sue Taylor Parker and Kathleen Rita Gibson, eds.), *supra* note 4, at 525–526.

59. *Id.* at H. Lyn Miles *et al., supra* note 3, at 308–316.

60. Josep Call and Michael Tomasello, "Production and comprehension of referential pointing by orangutans *(Pongo pygmaeus),* 108 *Journal of Comparative Psychology* 307, 315 (1994).

61. *Id.* at 308–316.

62. Shoji Itakura and Masayuki Tanaka, "Use of experimenter-given cues during object-choice tasks by chimpanzees *(Pan troglodytes),* an orang-utan *(Pongo pygmaeus)* and human infants *(Homo sapiens),* 112 *Journal of Comparative Psychology* 119, 122–123 (1998). *But see* Michael Tomasello, Josep Call, and Andrea Gluckman, "Comprehension of novel communication signs by apes and human children," 68 *Child Development* 1067 (1997).

63. Personal communication from Rob Shumaker (undated in 2001).

64. Kim A. Bard, "'Social tool use' by free-ranging orangutans: A Piagetian and developmental perspective on the manipulation of an animate object," in Sue Taylor Parker and Kathleen Rita Gibson, eds., *supra* note 4, at 358–359.

65. *Id.* at 368–369.

66. Shoji Itakura, "An exploratory study of gaze-monitoring in nonhuman primates," 38 *Japanese Psychological Research* 174, (1996).

67. H. Lyn White Miles (Sue Taylor Parker and Kathleen Rita Gibson, eds.), *supra* note 4, at 525, 530, 535.

68. Juan Carlos Gómez, "Ostensive behavior in great apes: The role of eye contact," in Anne E. Russon *et al.,* eds., *supra* note 3, at 131, 140; Juan Carlos Gómez, "Non-human primate theories of (nonhuman primate) minds: Some issues concerning the origins of mindreading," in *Theories of Theories of Minds* 330, 338–340 (Peter Carruthers and Peter K. Smith, eds., Cambridge University Press 1996); J. C. Gómez and P. Teixidor, "Theory of mind in an orangutan: A nonverbal test of false-belief appreciation?" Paper presented at the XIV Congress of the International Primatological Society, Strasbourg, France (August 16–21, 1992).

69. Personal communication from Robert W. Shumaker (undated in 2001).

70. Robert W. Shumaker, "Observational learning in orangutans," undated master's thesis, George Mason University.

71. Personal communication from Rob W. Shumaker (undated in 2001).

72. *Id.*

73. Personal communication from Daniel Shillito (undated in 2000).

74. H. Lyn Miles *et al., supra* note 3, at 286–289, 292–293.

75. *Id.* at 278, 283, 284.

76. H. Lyn White Miles (Sue Taylor Parker and Kathleen Rita Gibson, eds.), *supra* note 4, at 513, 534.

77. Sue Taylor Parker and Michael L. McKinney, *Origins of Intelligence: The Evolution of Cognitive Development in Monkeys, Apes, and Humans* 133 (John Hopkins University Press 1999).

78. Personal communication from Rob W. Shumaker (undated in 2001); H. Lyn Miles *et al., supra* note 3, at 290–291, 293.

79. Personal communication from H. Lyn Miles, dated November 15, 2000.

80. Rob W. Shumaker, *supra* note 70. Michael Tomasello and Josep Call, *supra* note 24, at 353 (more conservatively and skeptically); Josep Call, "Levels of imitation and cognitive mechanisms in orangutans," in Sue Taylor Parker *et al.,* eds., *supra* note 4, at 316, 327–339; Josep Call and Michael Tomasello, "The use of social information in the problem-solving of orangutans *(Pongo pygmaeus)* and human children *(Homo sapiens),*" 109 *Journal of Comparative Psychology* 308, 316, 318–319 (1995).

81. Anne E. Russon, *supra* note 5, at 76.

82. Anne E. Russon, "Orangutans' imitation of tool use: A cognitive interpretation," in Sue Taylor Parker *et al.,* eds., *supra* note 4, at 117, 117.

83. Both examples are described in Richard W. Byrne and Anne E. Russon, "Learning by imitation: A hierarchical approach," in 21 *Behavioral and Brain Sciences* 667, 669–680 (1998). The attempted fire-lighting is also described, with photographs, in Anne E. Russon, *supra* note 5, at 76–80, and in Anne E. Russon and Birute M. F. Galdikas, "Imitation in free-ranging rehabilitant orangutans *(Pongo pygmaeus),* 107 *Journal of Comparative Psychology* 147, 151, 153–154 (1993).

84. Anne E. Russon, *supra* note 82, at 134–135; Anne E. Russon, *supra* note 5, at 79–81; Anne E. Russon and Birute M. F. Galdikas, *supra* note 83, at 151.

85. Anne E. Russon, *supra* note 5, at 81.

86. H. Lyn Miles, "How can I tell a lie? Apes, language, and the problem of deception," in *Deception: Perspectives on Human and Nonhuman Deceit* 245, 253–258 (Robert W. Mitchell and Nicholas S. Thompson, eds., SUNY Press 1986).

87. H. Lyn Miles, *supra* note 86, at 262–264. *See* Robert W. Mitchell, "A framework for discussing deception," in *Deception: Perspectives on Human and Nonhuman Deceit* 3, 21–28 (Robert W. Mitchell and Nicholas S. Thompson, eds., SUNY Press 1986).

88. *Id.* at 257, 263.

89. H. Lyn White Miles (Sue Taylor Parker and Kathleen Rita Gibson, eds.), *supra* note 4, at 529; H. Lyn Miles, *supra* note 86, at 263.

90. Personal communication from H. Lyn Miles, dated November 15, 2000.

91. *Id.*

92. Personal communication from Sue Taylor Parker, dated November 15, 1999 (Step 7); Sue Taylor Parker and Michael L. McKinney, *supra* note 77, at 127, 131, 134. In her personal communication, Parker said she hesitated to label Chantek's role-reversal as Step 7 because it happened only once. But Lyn Miles reports it happened multiple times, personal communication from H. Lyn Miles, dated November 15, 2000.

93. Personal communication from H. Lyn Miles, dated November 15, 2000.

94. Anne E. Russon, "Pretending in free-ranging rehabilitant orangutans" and "Seeing with the mind's eye: Eye covering play in orangutans and Japanese macaques" in *Pretending and Imagination in Animals and Children* 241, 242 (Robert W. Mitchell, ed., Cambridge University Press 2002). Some scientists insist that pretend representations be symbolic. Russon disagrees, *id.*

95. Sue Taylor Parker and Michael L. McKinney, *supra* note 77, at 124–126.

96. *Id.;* H. Lyn Miles *et al., supra* note 86, at 284–285.

97. Personal communication from H. Lyn Miles, dated November 15, 2000.

98. Personal communication from Sue Taylor Parker, dated November 15, 1999; Sue Taylor Parker and Michael L. McKinney, *supra* note 78, at 125, 134.

99. Anne E. Russon, *supra* note 94, at 254.

100. *Id.*

101. Personal communication from Anne Russon, dated January 28, 2001.

102. Personal communication from Anne Russon, dated January 28, 2001.

103. *Id.* at Anne E. Russon, "Deception in Orangutans," in Robert W. Mitchell, ed., *supra* note 94, at 229, 237. *See* Robert W. Mitchell, "Deception and concealment as strategic script violation in great apes and humans," in Sue Taylor Parker *et al.,* eds., *supra* note 4, at 295, 302, 304–305, for a list of reported orangutan deceptions.

104. Personal communication from Daniel Shillito (undated in 2000).

105. Kim A. Bard, "Intentional behavior and intentional communication in young free-ranging orangutans," 63 *Child Development* 1186, 1191–1194 (1992).

106. Gary L. Shapiro and Birute M. F. Galdikas, "Early sign performance in a free-ranging, adult orangutan," in Sue Taylor Parker *et al.,* eds., *supra* note 4, at 265, 265.

107. *Id.* at 266. For more information on Project Rinnie, *see* Gary L. Shapiro and Birute M. F. Galdikas, "Attentiveness of orangutans within the sign learning context," in R. D. Nadler *et al., The Neglected Ape* 199 (Plenus Press 1995); Gary L. Shapiro, "Sign acquisition in a home-reared/free-ranging orangutan: Comparisons with other signing apes," 3 *American Journal of Primatology* 121 (1982).

108. Gary L. Shapiro and Birute M. F. Galdikas, *supra* note 106, at 266.

109. *Id.* at 268–269.

110. *Id.*

111. *Id.* at 274.

112. Anne E. Russon, *supra* note 5, at 145.

113. H. Lyn White Miles (Sue Taylor Parker and Kathleen Rita Gibson, eds.), *supra* note 4, at 512; H. Lyn Miles (Judith de Luce and Hugh T. Wilder, eds.), *supra* note 4, at 48; H. Lyn Miles, "Symbolic communication with and by great apes," in Sue Taylor Parker *et al.,* eds., *supra* note 4, at 197, 202, 205–206.

114. H. Lyn Miles (Sue Taylor Parker *et al.,* eds.), *supra* note 93, at 202; H. Lyn White Miles (Sue Taylor Parker and Kathleen Rita Gibson, eds.), *supra* note 4, at 519.

115. H. Lyn Miles and Stephen E. Harper, "Chantek: The language ability of an enculturated orangutan *(Pongo pygmaeus),*" in *Proceedings of The International Conference on "Orangutans: The Neglected Ape,"* 209, 211 (Zoological Society of San Diego 1994); H. Lyn Miles, "Symbolic communication with and by great apes," *supra* note 113, at 203; H. Lyn White Miles (Sue Taylor Parker and Kathleen Rita Gibson, eds.), *supra* note 4, at 513; H. Lyn Miles (Judith de Luce and Hugh T. Wilder, eds.), *supra* note 4, at 49.

116. H. Lyn White Miles (Sue Taylor Parker and Kathleen Rita Gibson, eds.), *supra* note 4, at 513–514.

117. *Id.* at 535.

118. H. Lyn Miles and Stephen E. Harper, *supra* note 115, at 211; H. Lyn Miles, (Judith de Luce and Hugh T. Wilder, eds.), *supra* note 4, at 48.

119. H. Lyn Miles *et al., supra* note 3, at 283, 284; H. Lyn White Miles (Sue Taylor Parker and Kathleen Rita Gibson, eds.), *supra* note 4, at 513.

120. Francine Patterson and Eugene Linden, *The Education of Koko* 76 (Holt, Rinehart and Winston 1981).

121. John D. Bonvillian and Francine G. P. Patterson, "Early sign-language: Comparisons between children and gorillas," in Sue Taylor Parker *et al., supra* note 7, at 245–246; Francine Patterson and Eugene Linden, *supra* note 120, at 75–76; Francine G. Patterson, "The gestures of a gorilla: Language acquisition in another pongid," 5 *Brain and Language* 72, 76 (1978).

122. Francine G. P. Patterson and Ronald H. Cohn, "Language acquisition by a lowland gorilla: Koko's first ten years of vocabulary development," 41 *Word* 97, 104 (August 1990).

123. H. Lyn Miles (Judith de Luce and Hugh T. Wilder, eds.), *supra* note 4, at 49.

124. H. Lyn White Miles (Sue Taylor Parker and Kathleen Rita Gibson, eds.), *supra* note 4, at 517.

125. *Id.* at 515–516.

126. *Id.* at 516, 525.

127. H. Lyn White Miles (Sue Taylor Parker *et al.*, eds.), *supra* note 4, at 264–265.

128. *Id.* at 260.

129. H. Lyn White Miles (Sue Taylor Parker and Kathleen Rita Gibson, eds.), *supra* note 4, at 528; H. Lyn Miles (Judith de Luce and Hugh T. Wilder, eds.), *supra* note 4, at 55.

130. H. Lyn White Miles (Sue Taylor Parker and Kathleen Rita Gibson, eds.), *supra* note 4, at 528.

131. *Id.* at 535.

132. Personal communication from Robert W. Shumaker (undated in 2001); personal communication from Robert W. Shumaker, dated December 6, 2000; "The Orangutan Language Project at the National Zoo, Summer 2000 (video). The Web site of the Orangutan Language Project can be found at http://natzoo.si.edu/zooview/exhibits/thinktank/olp/olp.htm (visited December 2, 2000).

133. Conversation with Ben Beck on June 30, 2000, at "ThinkTank," National Zoo, Washington, D.C.

134. Gisela Kaplan and Lesley J. Rogers, *supra* note 16, at 107.

135. Dan Shillito, Rob Shumaker, Gordon Gallup, and Ben Beck (Ph.D. dissertation of Dan Shillito, in progress). The facts I relate are based upon a telephone conversation I had with Dan Shillito on May 23, 1999, a videotape that shows a sampling of Indah's trials, and personal communications from Dan Shillito, dated June 9 and July 5, 1999.

136. H. Lyn Miles *et al., supra* note 3, at 285.

137. H. Lyn White Miles (Sue Taylor Parker and Kathleen Rita Gibson, eds.), *supra* note 4, at 529.

138. *Id.* at 265.

139. Anne E. Russon, *supra* note 5, at 88.

140. Thanks to Anne Russon for reviewing my description of what we saw on the videos on October 25 and November 6, 2000.

Chapter 11 Koko

1. T. Meehan, "Disease concerns in lowland gorillas," in *Management of Gorillas in Captivity* 153 (J. Ogden and D. Wharton, eds., The Gorilla Species Survival Plan and the Atlanta/Fulton County Zoo 1997).

2. "Ape said to have used sign language dies at 27," *New York Times*, at p. 18 (April 21, 2000). To learn about Michael, *see* "Memorial issue," *Gorilla* (2000) (*Gorilla* is the non-peer-reviewed publication of The Gorilla Foundation).

3. *Gorilla, supra* note 2, at 6.

4. Williams is an actor and comedian, Sting an actor and popular singer.

5. *Playboy* 183 (December 1986).

6. Edward Klima and Ursula Bellugi, *The Signs of Language* 189 (Harvard University Press 1979).

7. Morris Goodman *et al.*, "Toward a phylogenetic classification of primates based on DNA evidence complemented by fossil evidence," 9(3) *Molecular Phylogenetics and Evolution* 585, 594 (1998).

8. John D. Bonvillian and Francine G. P. Patterson, "Sign language acquisition and the development of meaning in a lowland gorilla," *The Problem of Meaning: Behavioral and Cognitive Processes* 181, 209 (1997).

9. Francine G. P. Patterson and Ronald H. Cohn, "Language acquisition by a lowland gorilla: Koko's first ten years of vocabulary development," 41 *Word* 97, 116 (August 1990); Francine Patterson and Eugene Linden, *The Education of Koko* 124–128 (Holt, Rinehart and Winston 1981).

10. Juan Carlos Gómez, "Non-human primate theories of (non-human primate) minds: Some issues concerning the origins of mind-reading," in *Theories of Theories of Mind* 330, 333, 335 (Peter Carruthers and Peter K. Smith, eds., Cambridge University Press 1996).

11. *Id.*

12. *Id.* at 339.

13. *Id.* at 339–340.

14. *Id.* at 343.

15. *Id.* at 335.

16. Juan Carlos Gómez, "Ostensive behavior in great apes: The role of eye contact," in *Reaching Into Thought: The Minds of the Great Apes* 131, 138, 142 (Anne E. Russon *et al.*, eds., Cambridge University Press 1996).

17. *Id.* at 143–144.

18. *Id.* at 145–147; Juan Carlos Gómez, "Mutual awareness in primate communication: A Gricean approach," in *Self-Awareness in Animals and Humans: Developmental Perspectives* 61, 71–75 (Sue Taylor Parker *et al.*, eds., Cambridge University Press 1994); Juan Carlos Gómez, "Visual behavior as a window for reading the mind of others in primates," in *Natural Theories of Mind: Evolution, Development and Simulation of Everyday Mindreading* 195, 201, 203, 206 (Andrew Whiten, ed., Basil Blackwell 1991). *See* Andrew Whiten, "When does smart behavior-reading become mind-reading?" in Peter Carruthers and Peter K. Smith, eds., *supra* note 10, at 277, 279–280.

19. Personal communication from Joanne Tanner, dated March 20, 2001.

20. This definition blends two definitions of "gesture" that Tanner and Byrne gave. *See* Joanne E. Tanner and Richard W. Byrne, "The development of spontaneous gestural communication in a group of zoo-living lowland gorillas," in *The Mentalities of Gorillas and Orangutans* 211, 216, 228–229 (Sue Taylor Parker *et al.*, eds., Cambridge University Press 1999); Joanne E. Tanner and Richard W. Byrne, "Representation of action through iconic gesture in a captive lowland gorilla," 37 *Current Anthropology* 162, 163–169 (1996).

Captive bonobos use iconic gestures, too, *see* E. S. Savage-Rumbaugh *et al.*, "Spontaneous gestural communication in the pygmy chimpanzee *(Pan paniscus),*" in *Progress in Ape Research* 97 (Geoffrey Bourne, ed., Academic Press 1977).

21. Joanne E. Tanner and Richard W. Byrne (in Sue Taylor Parker *et al.*, eds.) *supra* note 20, at 217–220; Joanne E. Tanner and Richard W. Byrne *(Current Anthropology), supra* note 20, at 166, 167.

22. Joanne E. Tanner and Richard W. Byrne (in Sue Taylor Parker *et al.*, eds.), *supra* note 20, at 234–235.

23. *Id.* at 233–234.

24. Francine G. Patterson, "Gorilla language acquisition," 17 *National Geographic Society Research Reports* 677, 681 (1976).

25. John D. Bonvillian and Francine G. Patterson, s*upra* note 8, at 240.

26. *See* Francine Patterson *et al.*, "Pragmatic analysis of gorilla utterances," 12 *Journal of Pragmatics* 35 (1988) (compares Koko and children at the one word stage of language development).

27. Francine Patterson and Eugene Linden, *supra* note 9, at 30.

28. Francine Patterson and Eugene Linden, *supra* note 9, at 92, 107; Francine G. Patterson, *supra* note 24, at 685–686.

29. John D. Bonvillian and Francine G. Patterson, *supra* note 8, at 248–250, 258, 260.

30. *Id.* at 244; Francine G. Patterson, "The gestures of a gorilla," 5 *Brain and Language* 72, 77–79 (1978).

31. Francine G. P. Patterson and Ronald H. Cohn, *supra* note 9, at 102.

32. Francine G. Patterson, *supra* note 30, at 78.

33. Francine G. P. Patterson and Ronald H. Cohn, "Language acquisition in a lowland gorilla: Koko's first ten years of vocabulary development," 14 *Gorilla* 4 (December 1990); Francine G. P. Patterson and Candace L. Holts, "A review of Project Koko: Language acquisition by two lowland gorillas, Part 1," 12(2) *Gorilla* 2 (June 1989); Francine Patterson and Eugene Linden, *supra* note 9, at 84. A list of the words Koko acquired in the first ten years of Project Koko may be found at Francine G. P. Patterson and Ronald H. Cohn, *supra* note 9, at 133–140.

34. Francine G. P. Patterson and Ronald H. Cohn, *supra* note 9, at 106; Francine G. Patterson, *supra* note 24, at 687.

35. Francine Patterson, "Why Koko (and Michael) *can* talk," 6(2) *Gorilla* 1, 1 (June 1983).

36. Francine Patterson, "Conversations with a gorilla," 154 *National Geographic* 438, 456 (October 1978).

37. Francine Patterson and Wendy Gordon, "The case for the personhood of gorillas," in *The Great Ape Project: Equality Beyond Humanity* 63 (Paola Cavalieri and Peter Singer, eds., Fourth Estate 1993).

38. Francine Patterson and Eugene Linden, *supra* note 9, at 117.

39. Personal communication from Penny Patterson (undated in 2001).

40. Francine G. Patterson, *supra* note 30, at 91.

41. Francine G. Patterson, *supra* note 24, at 688.

42. Francine Patterson and Wendy Gordon, *supra* note 37, at 59; Ronald Cohn and Francine Patterson, "Koko speaks English using Apple Computer," 15 *Gorilla* 2 (December 1991); Francine Patterson and Eugene Linden, *supra* note 9, at 105.

43. *See* Joanne Tanner and Wendy Gordon, "Reading with Koko: The word *duck* rhymes with *chuck*," 15(1) *Gorilla* 7–8 (December 1991); Joanne Tanner, "Koko: Reading report, 8(2) *Gorilla* 4–6 (June 1985); Anne Longman, "Learning to read," 7(2) *Gorilla* 4 (June 1984); Joanne Tanner, "Comment about Koko's reading," 7(1) *Gorilla* 7 (December 1983); Joanne Tanner *et al.,* "Koko makes the reading connection," 7(1) *Gorilla* 6–7 (December 1983); Anne Longman, "Reading project," 6(2) *Gorilla* 6 (June 1983).

44. Francine Patterson and Eugene Linden, *supra* note 9, at 100–105; Francine G. Patterson, *supra* note 24, at 684.

45. Francine G. Patterson, *supra* note 30, at 83–84.

46. Personal communication from Penny Patterson, dated January 9, 2001. Patterson has a photograph of Koko doing this.

47. Francine G. Patterson, *supra* note 30, at 85–86.

48. John D. Bonvillian and Francine G. P. Patterson, "Early sign language acquisition in children and gorilla: Vocabulary content and sign iconicity," 13 *First Language* 315, 318 (1993).

49. *Id.* at 333.

50. "A conversation with Koko," produced for *Nature* (1999), at 31:25; Joanne E. Tanner and Richard W. Byrne (Sue Taylor Parker *et al.,* eds.), *supra* note 20, at 210, 234; Francine Patterson and Wendy Gordon, *supra* note 37, at 64–65.

51. H. Lyn Miles, "Symbolic communication with and by great apes," in Sue Taylor Parker *et al.,* eds., *supra* note 20, at 197, 201.

52. Francine G. Patterson, *supra* note 24, at 681.

53. *Id.* at 682.

54. Francine G. Patterson, *supra* note 30, at 87. For "TOOTHBRUSH," *see* "A Conversation with Koko," *supra* note 50, at 20:27.

55. Francine Patterson and Eugene Linden, *supra* note 9, at 137.

56. Suzanne Chevalier-Skolnikoff, "The Clever Hans phenomenon, cueing, and ape signing: A Piagetian analysis of methods for instructing animals," in *The Clever Hans Phenomenon: Communications With Horses, Whales, Apes, and People* 60, 80 (Thomas A. Sebeok and Robert Rosenthal, eds., The New York Academy of Sciences 1981).

57. Francine Patterson, *supra* note 36, at 459. *See* Francine G. Patterson, *supra* note 24, at 697–699.

58. Francine Patterson and Eugene Linden, *supra* note 9, at 100–105.

59. Francine Patterson, *supra* note 35, at 1; Francine Patterson and Wendy Gordon, *supra* note 37, at 65; Francine Patterson and Eugene Linden, *supra* note 9, at 146; Francine G. Patterson, *supra* note 30, at 88; Francine G. Patterson, *supra* note 28, at 693. For "SCRATCH COMB," *see* " A Conversation with Koko," *supra* note 50, at 32:07.

60. Francine Patterson and Wendy Gordon, *supra* note 3, at 65; Suzanne Chevalier-Skolnikoff, *supra* note 60, at 81–82.

61. Francine G. Patterson, *supra* note 24, at 693–694.

62. Francine Patterson, *supra* note 36, at 456.

63. "A Conversation with Koko," *supra* note 50, at 8:45 and 29:10.

64. Personal communication from Penny Patterson, dated January 9, 2001; Francine Patterson and Eugene Linden, *supra* note 9, at 77, 80, 205–207. *See* Francine G. Patterson, *supra* note 24, at 694–696.

65. Suzanne Chevalier-Skolnikoff, *supra* note 60, at 74.

66. *E.g.*, H. Lyn Miles, "How can I tell a lie? Apes, language, and the problem of deception," in *Deception: Perspectives on Human and Nonhuman Deceit* 245, 249–252 (Robert W. Mitchell and Nicholas S. Thompson, eds., State University of New York Press 1985); Thomas Sebeok, "Looking in the destination for what should have been sought at the source," in *Speaking of Apes* 407, 421 (Thomas A. Sebeok and Ann Umiker Sebeok, eds., Plenum Press 1980). *But see* Richard W. Byrne, *The Thinking Ape: Evolutionary Origins of Intelligence* 139 (Oxford University Press 1995).

67. Sarah T. Boysen, "Tool use in captive gorillas," in Sue Taylor Parker *et al.*, eds., *supra* note 20, at 178, 180. Oh please, I don't know Jennifer Lopez.

68. Juan C. Gómez, "Development of sensorimotor intelligence in infant gorillas: The manipulation of objects in problem-solving and exploration," in Sue Taylor Parker *et al.*, eds., *supra* note 20, at 160, 162–163.

69. B. Fontaine *et al.*, "Observations of spontaneous tool making and tool use in a captive group of western lowland gorillas *(Gorilla gorilla gorilla)*," 65 *Folia Primatologia* 219–223 (1995).

70. Giovanna Spinozzi and Patricia Poti, "Causality I: The support problem," in Francesco Antinucci, ed., *Cognitive Structure and Development in Nonhuman Primates* 89, 113, 116–119 (Lawrence Erlbaum 1989).

71. Sarah T. Boysen, *supra* note 67, at 184.

72. Benjamin Beck, *Animal Tool Behavior: The Use and Manufacture of Tools by Animal*s 10 (Garland STPM Press 1980).

73. Sue T. Parker *et al.*, "A survey of tool use in zoo gorillas," in Sue Taylor Parker *et al.*, eds., *supra* note 20, at 188–192.

74. *Id.* at 190–191.

75. *Id.* at 192. *See* Sue Taylor Parker and Robert W. Mitchell, "The mentalities of gorillas and orangutans in phylogenetic perspective," in Sue Taylor Parker *et al.*, eds., *supra* note 20, at 397, 403, 404 (lists kinds of simple and advanced tool uses observed in gorillas).

76. Juan C. Gómez, *supra* note 68, at 165.

77. *Id.* at 168. Another gorilla used a towel as a rake, Francesco Natale, "Stage 5 object-concept," in Francesco Antinucci, ed., *supra* note 70, at 93–95 (Lawrence Erlbaum 1989).

78. Juan Carlos Gómez, *supra* note 10, at 338–339.

79. *Id.* at 341. *See* Juan C. Gómez, *supra* note 68, at 165–166.

80. *Id.* at 169–175.

81. Personal communication from Penny Patterson, dated January 10, 2001; Suzanne Chevalier-Skolnikoff, "A Piagetian model for describing and comparing the socialization of monkey, ape, and human infants," in *Primate Biosocial Development* 159, 168 (S. Chevalier-Skolnikoff and F. Poirier, eds., Garland 1977). Figure 9 shows Koko engaging in a tertiary circular reaction as she tried to remove honey from a jar, *id.* at 180.

82. Juan C. Gómez, *supra* note 68, at 175; Juan Carlos Gómez, *supra* note 10, at 348–349, 352; Juan Carlos Gómez *(Natural Theories of Mind)*, *supra* note 18, at 197–200.

83. Juan Carlos Gómez, *supra* note 10, at 344.

84. *Id.* at 342.

85. Juan Carlos Gómez, *supra* note 18 *(Self-awareness in Animals and Humans)*, at 70.

86. "A conversation with Koko," *supra* note 50, at 00.16–00.27

87. Francine G. P. Patterson and Ronald H. Cohn *(Gorilla), supra* note 33, at 4; Francine Patterson and Joanne Tanner, "Mirror behavior and self-concept in the lowland gorilla," 14(1) *Gorilla* 2 (December 1990).

88. Francine Patterson, "Self-awareness in the gorilla Koko," 14(2) *Gorilla* 2, 3 (June 1991).

89. Francine Patterson and Joanne Tanner, *supra* note 87, at 2.

90. Francine Patterson, *supra* note 88, at 3–4; Daniel Hart and Mary Pat Karmel, "Self-awareness and self-knowledge in humans, apes, and monkeys," in Anne Russon *et al.*, eds., *supra* note 16, at 325, 332–333.

91. Karyl B. Swartz *et al.*, "Comparative aspects of mirror self-recognition in great apes," in Sue Taylor Parker *et al., supra* note 20, at 281, 287–289.

92. Conversation with Penny Patterson, November 19, 2000; Francine Patterson, *supra* note 88, at 2.

93. Francine Patterson, *supra* note 88, at 2.

94. Francine G. P. Patterson and Ronald H. Cohn, "Self-recognition and self-awareness in lowland gorillas," in Sue Taylor Parker *et al.,* eds., *supra* note 18, at 273–278; Francine Patterson and Joanne Tanner, *supra* note 87, at 2.

95. For a photograph of Koko twisting her face about, the better to see a spot on her left cheek in a mirror, *see* Richard Byrne, *supra* note 66, at 116, or Francine Patterson and Joanne Tanner, *supra* note 87, at 2. Koko admires her face made up with chalk, at Francine Patterson, *supra* note 36, at 439.

96. Lindsey E. Law and Andrew J. Lock, "Do gorillas recognize themselves on television?" in Sue Taylor Parker *et al.*, eds., *supra* note 18, at 308, 308–311.

97. Karyl B. Swartz and Sian Evans, "Social and cognitive factors in chimpanzee and gorilla mirror behavior and self-recognition," in Sue Taylor Parker *et al.,* eds., *supra* note 18, at 189, 201–202.

98. Sue Taylor Parker, "Incipient mirror self-recognition in zoo gorillas and chimpanzees," in Sue Taylor Parker *et al.,* eds., *supra* note 18, at 301, 301, 305.

99. *The Oxford English Dictionary* 543, "ape" (defs. 2b, 3a, 3b, 4).

100. *Id.* at 114–119; Richard Byrne, *supra* note 66, at 68–71, 75.

101. Richard Byrne, *supra* note 66, at 68–69 (Byrne erroneously called this emulation "stimulus enhancement," personal communication from Richard Byrne, dated November 8, 2000); Richard W. Byrne and Anne E. Russon, "Learning by imitation: A hierarchical approach," 21 *Behavioural and Brain Sciences* 667, 675 (1998).

102. Richard W. Byrne, *supra* note 66, at 118. *See* Richard W. Byrne and Jennifer M. E. Byrne, "Complex leaf-gathering skills of mountain gorillas *(Gorilla g. berengei)* variability and standardization," 31 *American Journal of Primatology* 241, 259 (1993).

103. T. S. Stoinski *et al.,* "Imitative learning of food-processing techniques in captive western lowland gorillas," *Journal of Comparative Psychology* 272, 280 (March 2002).

104. Francine Patterson and Eugene Linden, *supra* note 9, at 27, 32.

105. Francine G. Patterson, *supra* note 30, at 81–82; Francine Patterson and Eugene Linden, *supra* note 9, at 32.

106. Francine Patterson and Eugene Linden, *supra* note 9, at 40.

107. *Id.* at 38.

108. *Id.* at 104. Francine G. Patterson, supra note 24, at 686–687.

109. Francine Patterson and Eugene Linden, *supra* note 9, at 142–143.

110. Richard W. Byrne, *supra* note 66, at 139.

111. Francine G. P. Patterson and Ronald H. Cohn, *supra* note 94, at 284.

112. Robert W. Mitchell, "Deception and concealment as strategic script violation in great apes and humans," in Sue Taylor Parker *et al.*, eds., *supra* note 20, at 295, 295, 298, 300–301.

113. *Id.* at 302, 303, 304–309.

114. Joanne E. Tanner and Richard W. Byrne, *supra* note 20, at 229; Richard W. Byrne, *supra* note 66, at 120, 121, 123.

115. Joanne E. Tanner and Richard W. Byrne, *supra* note 20, at 231.

116. Personal communication from Penny Patterson (undated 2001); Joanne E. Tanner and Richard W. Byrne, "Concealing facial evidence of mood: Perspective-taking in a captive gorilla?" 34 *Primates* 451, 451–456 (October 1993). *See* Michael Tomasello and Josep Call, *Primate Cognition* 236 (Oxford University Press 1997); Robert W. Mitchell, "Deception and hiding in captive lowland gorillas *(Gorilla gorilla gorilla),"* 32(4) *Primates* 523 (1991).

117. Conversation with Dan Shillito, January 26, 2001.

118. Frans de Waal, *Good Natured: The Origins of Right and Wrong in Humans and Other Animals* 44 (Harvard University Press 1996). The story was first related by Heini Hediger in *Studies in the Psychology and Behavior of Animals in Zoos and Circuses* (Butterworth 1955).

119. "Encounters with deception," 20(1) *Gorilla* 5 (Winter/Spring 1997); Francine G. P. Patterson and Ronald H. Cohn, *supra* note 94, at 283.

120. "Encounters with deception," 20(1) *Gorilla* 5 (Winter/Spring 1997).

121. Francine Patterson and Eugene Linden, *supra* note 9, at 152.

122. Francine G. P. Patterson and Ronald H. Cohn, *supra* note 18, at 283; Francine Patterson, *supra* note 36, at 458.

123. Joanne E. Tanner and Richard W. Byrne, *supra* note 116, at 451–456.

124. John H. Flavell *et al.*, "Young children's knowledge about visual perception: Further evidence for the level 1–level 2 distinction," 17 *Developmental Psychology* 99–103 (1981).

125. Juan Carlos Gómez, *supra* note 16, at 145–147; Juan Carlos Gómez, *supra* note 18, at 201, 203, 206. *See* Andrew Whiten, "When does smart behavior-reading become mind-reading?" in Peter Carruthers and Peter K. Smith, eds., *supra* note 10, at 277, 279–280.

126. Andrew Whiten, "Parental encouragement in *Gorilla* in comparative social perspective: Implications for social cognition and the evolution of teaching," in Sue Taylor Parker *et al.,* eds., *supra* note 18, at 342, 356–357.

127. *Id.* at 343–344.

128. *But see* Timothy M. Caro and Marc D. Hauser, "Is there teaching in nonhuman animals?" 67 *The Quarterly Review of Biology* 151, 153 (1992) (adopt a definition of teaching that sidesteps the question of whether a teacher intends to educate and focuses on whether the pupil learns).

129. Andrew Whiten, *supra* note 126, at 345–347.

130. *Id.* at 361–362.

131. *Id.* at 349, 351, 362–363.

132. Suzanne Chevalier-Skolnikoff, *supra* note 56, at 84.

133. John D. Bonvillian and Francine G. P. Patterson, *supra* note 8, at 211, 214.

134. Conversation with Penny Patterson, November 19, 2000; Francine Patterson, *supra* note 99, at 2.

135. Francine G. P. Patterson *et al.*, "Story-telling by two Western lowland gorillas," in Proceedings of the Conference of Emotion, Kyoto, Japan (1999) (in press).

136. Michael Tomasello and Josep Call, *supra* note 116, at 40.

137. Francesco Natale *et al.*, "Stage 6 object concept in nonhuman primate cognition: A comparison between gorilla *(Gorilla gorilla gorilla)* and Japanese macaque *(Macaca fuscata)*, 100 *Journal of Comparative Psychology* 335, 339 (1986). *See* Josep Call and Michael Tomasello, "The effect of humans on the cognitive development of apes," in Anne E. Russon *et al.*, eds., *supra* note 16, at 370, 373 ("in the laboratory any number of chimpanzees and gorillas who have had many different types of experience with humans, some very limited, have solved all of the object permanence problems presented to them, including state 6 invisible displacements").

138. Personal communication from Penny Patterson (undated in 2001); John D. Bonvillian and Francine G. P. Patterson, *supra* note 8, at 207–208.

Chapter 12 Legal Rights for Nonhuman Animals

1. *Skinner v. Oklahoma,* 316 U.S. 535, 542 (1942).

2. *Id.* at 541.

3. *Id.* at 540.

4. *Id.* at 542.

5. *See* Peter Westen, *Speaking of Equality: An Analysis of the Rhetorical Force of "Equality" in Moral and Legal Discourse* xiv, note 2 (Princeton University Press 1990).

6. *Id.*

7. *E.g.,* Eric Rakowski, *Equal Justice* (Oxford University Press 1991); Peter Westen, *supra* note 5; Douglas Rae, *Equalities* (Harvard University Press 1981).

8. *The Belgian Linguistics Case,* A6., at para. 34 (Eur. Ct. Hum. R. 1968) (final judgment).

9. *Darby v. Sweden,* A87, at para. 31; *The Belgian Linguistics Case, supra* note 8, at 34.

10. *See, e.g., Romer v. Evans,* 517 U.S. 620, 634–635 (1996); *Rinaldi v. Yeager,* 384 U.S. 305, 308–309 (1966).

11. Race (*Korematsu v. United States,* 323 U.S. 214) (1944); *East African Asians Cases;* 3 EHRR 76 (1973); sex (*United States v. Virginia,* 518 U.S. 515, 531–534) (1996); *Abdulaziz, Cabales, and Balkandali v. UK,* A.94, at para. 78 (Eur. Ct. Hum. R. 1985); illegitimacy (*Trimble v. Gordon,* 430 U.S. 762) (1977); *Marckx v. Belgium,* A19, at para. 48 (Eur. Ct. Hum. R. 1975). *Compare Weinburger v. Salfi,* 422 U.S. 749, 785 (1975) *with Cleveland Board of Education v. LaFleur,* 414 U.S. 632 (1974); *Stanley v. Illinois,* 405 U.S. 645 (1972).

12. *United States v. Virginia, supra* note 11, at 542, quoting *Mississippi University for Women v. Hogan,* 458 U.S. 718, 725 (1985) (emphasis added). The Court also said that "state actors may not rely on 'overbroad generalizations' to make 'judgments about people that are likely to . . . perpetrate historical patterns of discrimination,'" *id.* at 533, or "to create or perpetrate the legal, social, and economic inferiority of women," *id.* at 534.

13. J. G. Randall, *Lincoln and the South* 33 (Louisiana State University 1946); The Fifth Joint Debate at Galesburg, October 7, 1858, in *The Lincoln-Douglas Debates* 246 (Harold Holzer, ed., HarperCollins 1993); The Sixth Joint Debate at Quincy, October 13, 1858, in *id.* at 300.

14. The Fifth Joint Debate at Galesburg, October 7, 1858, *supra* note 13, at 248.

15. David Herbert Donald, *Lincoln* 221 (Simon & Schuster 1995); The Sixth Joint Debate at Quincy, October 13, 1858, in Harold Holzer, ed., *supra* note 13, at 290; David Zarefsky, *Lincoln Douglas and Slavery: In the Crucible of Public Debate* 149 (University of Chicago Press 1990).

16. David M. Potter, *The Impending Crisis: 1848–1861* 346 (Harper & Row 1976).

17. The Fourth Joint Debate at Charleston, September 18, 1858, in Harold Holzer, ed., *supra* note 13, at 189.

18. The Sixth Joint Debate at Quincy, October 13, 1858, in Harold Holzer, ed., *supra* note 13, at 285.

19. David Zarefsky, *supra* note 15, at 243.

20. *Id.* at 34–35, 60–61, 193–194; Garry Wills, *Lincoln at Gettysburg: The Words that Remade America* 97 (Simon & Schuster 1992). *See* The First Joint Debate at Ottawa, August 21, 1858, in Harold Holzer, ed., *supra* note 13, at 63; The Fourth Joint Debate at Charleston, September 18, 1858, in *id.* at 189; The Sixth Joint Debate at Quincy, October 13, 1858, in *id.* at 284.

21. Excerpt from presentation of Colin Blakemore to the Fifth International Congress on Bioethics, London, September 23, 2000, reported in Catherine Pepinster, *Independent Digital (UK),* September 24, 2000.

22. Steven M. Wise, *Rattling the Cage: Toward Legal Rights for Animals* 244–245 (Perseus Publishing 2000).

23. La. Rev. Stat. Ann. Sec. 9:123 (1990).

24. La. Rev. Stat. Ann. Sec. 9:124 and 9:126 (1990).

25. Daniel Dombrowski, *Babies and Beasts: The Argument from Marginal Cases* 14, 21, 22, 23, 26, 34, 49–50, 52, 54, 63, 76, 77, 90, 92, 106, 116, 127, 152, 155, 159, 167, 168, 169, 182–183 (University of Illinois Press 1997). *See, e.g.,* Carl Cohen, "The Case for the Use of Animals in Biomedical Research," 315 *New England Journal of Medicine* 865, 867 (1986); Richard A. Epstein, "The Next Rights Revolution?" *National Review* 44, 45 (November 8, 2000).

26. Gisela Kaplan and Lesley J. Rogers, *The Orangutans: Their Evolution, Behavior and Future* 157 (Perseus Publishing 2000).

27. Hugh Thomas, *The Slave Trade: The Story of the Atlantic Slave Trade 1440–1870* 536 (Simon & Schuster 1997), quoting Edward Fitzmaurice, *Life of William, Earl of Shelburne* (1875).

28. Ira Berlin, "Some are more equal," *New York Times Book Review* 25 (September 9, 2001).

29. David Brion Davis, "The enduring legacy of the South's Civil War victory," *New York Times* at WK 1, 6 (August 26, 2001).

INDEX

ABOUT THE AUTHOR

Steven M. Wise is a pioneer in the field of animal rights law. For over twenty years, he has split his time among litigating animal protection cases, teaching animal rights law courses at law schools, including Harvard, Vermont, and John Marshall, and writing and speaking about animal law. He has taken on virtually every type of animal-related case, from veterinary malpractice to defending the rights of condominium and cooperative owners to have companion animals. He has sued the U.S. Patent Office to stop it from issuing patents for genetically engineered animals, opened up animal care committees to public scrutiny, and represented animal rights activists charged with protest crimes or defamation stemming from protests. Wise's ultimate goal is to win "legal personhood" for at least some nonhuman animals. Author of the widely praised *Rattling The Cage: Towards Legal Rights for Animals,* he is the former president of the Animal Legal Defense Fund and current president of the Center for the Expansion of Fundamental Rights, Inc.

ABOUT THE CENTER FOR THE EXPANSION OF FUNDAMENTAL RIGHTS, INC.

Since 1995, the Center for the Expansion of Fundamental Rights, Inc. (CEFR) has been the only nonprofit, tax-exempt organization in the world with the primary purpose of obtaining fundamental legal rights for nonhuman animals, beginning with chimpanzees and bonobos. If you would like information about CEFR, please visist our website at www.cefr.org. If you would like to make a tax-deductible contribution, CEFR's address is 896 Beacon Street, Suite 303, Boston, Massachusetts 02215.